Cultures on Celluloid

Keith Reader

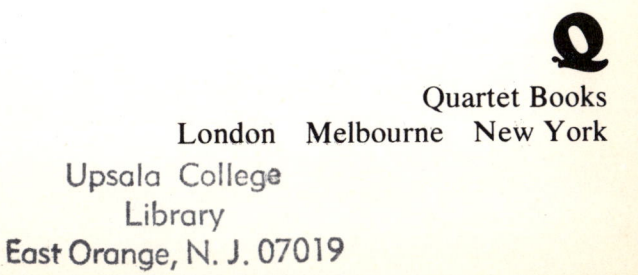

Quartet Books
London   Melbourne   New York

First published by Quartet Books Limited 1981
A member of the Namara Group
27 Goodge Street, London W1P 1FD

Copyright © 1981 by Keith Reader

ISBN 0 7043 2272 2

British Library Cataloguing in Publication Data

Reader, Keith
   Cultures on celluloid.
   1.   Moving pictures – Social aspects
   I. Title
   302.2′3   PN1995.9.S6

ISBN 0–7043–2272–2

Phototypeset by Amicon Print Ltd., Wallington, Surrey.
Printed and Bound Mansell (Bookbinders) Ltd., Witham, Essex

Cultures on Celluloid

Also by Keith Reader

The Cinema: A History

I dedicate this book to my father
– for more than he can ever know

Acknowledgements

My thanks go in particular to:
the libraries of the British Film Institute Information Department in London, the Bibliothèque de l'Arsenal and the Institut des Hautes Études Cinématographiques in Paris, and the Museum of Modern Art Film Archive in New York, whose staff gave me patient and kindly attention; the Service d'Accueil of the French Foreign Ministry, a grant from whom made it possible for me to spend a month in Paris researching the French section of the work in the summer of 1979; the Research Committee of Kingston Polytechnic, for granting me a term's study leave without which the book could not have been completed on time; Miss Jones of the National Film Theatre box-office, for help and kindness; Jill Forbes, Myron Kofman, Chris Cook, Ian Reader, Phillip Drummond and others for various forms of help, ideas, and stimulation; and above all to Eleonore, for endurance going far beyond the call of duty.

Contents

Introduction 1

Part One: The Organization – American Society in and through its Cinema 11

    The Silent Era 20
    The Early Sound Period and the Films of the Thirties 26
    Arcadia 29
    The Western and the Gangster Movie 41
    Gone with the Wind 50
    The War Years and *Film Noir* 53
    The Post-War Era 68
    The Sixties 83
    The Seventies – Violence and Vietnam 94

Part Two: History in the Writing – French Society through its Cinema 109

    The Early Days 114
    Jean Renoir 129
    The New Wave 138

Part Three: 'You will, Oscar, you will' – British Disavowal and Repression 153

    Kind Hearts and Coronets 162

Part Four: Inscrutability and Reproduction – Four Major Japanese Directors 171
    Ozu Yasujiro 177
    Kurosawa Akira 183
    Mizoguchi Kenji 188
    Oshima Nagisa 192

References 195
Glossary 198
Bibliography 200
Filmography 203
Index 210

Introduction

Perceptions of what different nationalities are 'like' rank among the most tenacious of cultural stereotypes. The indolent Spaniard, the amorous Italian, the diligent and humourless German are sufficiently hackneyed of their kind to raise a few indulgent laughs; yet it is no coincidence that the three nations selected were either at war with Britain or hostilely neutral towards her between 1939 and 1945, and this shows how even so apparently innocuous an example of national stereotyping can be pressed into the service of rivalries and antagonisms it may both precede and (in some form) survive.

Not that the present study is going to concern itself with the 'funny foreigner' mentality of the British seaside postcard. First and foremost, its concern will be to analyse the ways in which different national cultures depict *themselves* in and through their cinematic industries. Racism and prejudice, as the treatment of Indians in the American Western would alone go to show, are not thereby avoided; but they are often coopted into the service of some mythical larger unity (thus, the wholesale extermination of Indians in one Western after another is 'justified' by the threat they posed to the white man who built the 'America' that built the cinema in which the film's spectators are sitting). Hollywood in particular, but all the film cultures we shall look at in different ways, acted as propaganda for and endorsement of its country's 'way of life' within the country in question and abroad. This clearly entailed the selection and privileging of certain attitudes, activities, and even geographical regions over others. The negligible serious treatment given to Wales, Scotland and Ireland* in the British cinema; the expulsion of suspected 'Communists' (read Liberals) from the American industry in the McCarthy era, and the aggressive individualism that dominates the Western and gangster movie; the explicit mechanisms of political censorship in France; the pressures that led a Japanese film-maker such as Mizoguchi to couch many of his most vehement criticisms of twentieth-century Japanese society in the form of historical period dramas – these are a sample indication of how wide the range of the study could be, embracing as it does not only the films made but the structures and functioning of the various national industries.

What are the criteria of choice I have adopted? The book is intended in the first place for a British audience, and availability of individual films and other material in Britain has therefore been a paramount consideration. The industries of the United States, France, and Britain are certainly those whose products are most widely available here. In addition, these three nations (or cultural groupings) have exercised widespread cultural and ideological influence since they came into being, and more particularly through their filmic products since the invention of the cinema. I have attempted to consider the production of the United States in the light of the ideology (or ideologies) of market political and economic freedom, and that of France in the light of that nation's perception of itself as the first and archetypal modern nation-state – though as will soon become plain this certainly does not mean that all their films will be seen to reflect these attitudes. Pressures of space and the desire to try out slightly different methods of analysis led me to treat the British and Japanese industries in much shorter compass, focusing for the British industry on how a peculiarly perverse attitude towards sexuality – almost a desexualized prurience – leaves its mark even on so urbane a film as *Kind Hearts and Coronets*, and for the Japanese industry on how the extreme otherness of Japanese culture may be brought out (and has certainly been mediated into something like 'acceptability' for Western art-house audiences) through a consideration of the work of four major film-makers, one after another.

*I would add here that 'Ireland' refers to the island which (whatever one's views on its current national status) has for long been inextricably bound up with British history and economy.

Introduction

This leaves unanswered the manner in which material for inclusion in the two longest sections was chosen. Certain films effectively selected themselves: how to consider the importance of populism in American ideology without looking at *Citizen Kane*, how to understand the conflicts within French society immediately before the Second World War without *La Règle du Jeu*? Others were apparently deserving of attention, but less so than they appeared at first; thus, while consideration of Hollywood's treatment of the war in Vietnam would not have been complete without a reference (however sneering) to *The Green Berets*, that film's very grossness, not just on a propagandist level but also in its wholesale importation of the strategies and conflicts of the Western into a quite inapposite setting, meant that the amount of space that could profitably be devoted to it was relatively little. Others again proved interesting precisely because of the contradictions and inconsistencies they embodied. *Mildred Pierce*'s ambivalence towards its heroine (failed wife-and-mother, failed career woman, both, or neither?); the difficulty of assessing exactly how Jean Gabin's suicide at the end of *Le Jour se Lève* can be read in the context of 1930s France; the tantalizing absence of both politics and sexuality from the work of the Japanese director Ozu, none the less centrally concerned with that battleground of the two, the family – these are the kind of questions I have attempted to explore. That much of the interest of a film, as of any work of art or cultural document, can lie in what it does not say, or says confusingly or contradictorily, was a central conviction of mine when I started work on this book; it is more so than ever now.

A word needs to be said about what exactly is meant by the 'time' or 'period' of a film. How we read a society through its films has in part to do with how familiar we are with its history and culture, but we should never forget that these, as we perceive them in the cinema, are not genuine 'period pieces'. They are reconstructions (or stylizations, or even – as in Robin Hood and Merrie England costume-dramas – falsification) of a period to which direct access is obviously impossible. This has more than pedantic relevance. The manner in which Renoir reconstructs the Revolution of 1789 in *La Marseillaise*, or John Wayne the time of Davy Crockett in *The Alamo*, tells us a great deal more about the time of the films' making than about that in which they are set – a point which will become particularly important in our consideration of the French cinema.

But the history of (or in) a film does not come to a stop when the shooting finishes. The long history of censorship, and its constant proof that what is acceptable in one period may turn out to be 'pernicious' or 'subversive' in another, is proof enough of that. Films undergo a multiplicity of readings, rereadings, and revaluations from one period to the next. The theme of neutrality in that perennial favourite *Casablanca* would have been much more perceptible to an American audience of 1943 than it is likely to be today, just as the disrespect for authority that caused *Zéro de Conduite* to be banned will have lost a good deal of its force for a modern audience unacquainted with the exceptionally repressive structures of the French educational system of the time. It will be part of my concern here to bring into the foreground aspects such as these, which have become submerged precisely by the films' status as 'classics' – deserved, to be sure, but running the ever-present risk of removing them from that domain of history without which they lose much of their strength.

It can also be true that films dismissed as pot-boilers, or conversely restricted to a very small audience, on first release, subsequently acquire a very different status. This phenomenon, which I have dubbed 'secondary commercialization', is most obviously marked in the Hollywood cinema, where the work of low-budget directors such as Douglas Sirk and Samuel Fuller has been the focus of serious critical attention that has transplanted it from the bottom half of small-town cinema double bills to top-of-the-bill status at film festivals and art-houses. (Whether this is a 'good' or a 'bad' thing is a question I do not intend to attempt answering, largely because I see no sense in posing it in the first place). *Casablanca* has moved away from the commercial circuit (despite periodic reappearances), to the world of campus film-societies and late-night features, in which it is a major element. *Easy Rider* and *American Graffiti*, both shot on minimal budgets, became record-breaking commercial successes. Different readings here again constantly need to be borne in mind. To treat Sirk as a 'feminist' critic of the nuclear family, or *Easy Rider* as a major act of cultural rebellion against the status quo, is certainly to some extent valid. But Sirk was also a B-picture director operating within very precise constraints, and the cry of revolt in *Easy Rider* now appears to have

Introduction

certain very ambiguous overtones to it. Only attention to the circumstances of a film's making and initial release along with those of its later dissemination can take proper account of these variations and prevent the film from being treated as in some mysterious way outside history.

The French sociologist Pierre Sorlin says that 'because it is an item of goods, designed to be sold at a profit, every film puts into operation the rules of its own commercialization'.[1] As the phenomenon of secondary commercialization makes clear, this is not just a piece of vulgar economism reminding us that films must at least look as though they are going to make a profit in order to be made. The films of Sirk at the time of their release put into operation the rules of their own commercialization, among the most important of which was that they should not cost too much; hence the unrealistic cut-price sets and the recurring use of good but not enormously expensive actors. But these rules function in a very different way for a modern Sirk audience, at the Edinburgh Festival or Notting Hill's Electric Cinema Club. Here it is the ensemble playing and the revelatory stylization of the décor that make the films commercially viable, for we should not forget that even the most 'experimental' or 'artistic' of cinemas still needs to show a profit on its investment. There is, in other words, no single image of a nation or a culture that can be derived from any film about it, for that which emerges will be economically as well as historically different at different times.

For a concrete illustration of how all these factors might be combined into an analysis of an individual film, I shall look at John Ford's *The Searchers*, starring John Wayne, made in 1956, and regarded ever since as among the most important of both star's and director's work. The film's story is extremely simple, a variant of the quest theme that has been a narrative staple in the West ever since Homer's *Odyssey*. John Wayne/Ethan Edwards* sets off in search of his niece, Debbie (played by Natalie Wood), who has been abducted by the Comanche Indians responsible for the massacre of her parents and the destruction of their homestead. His journey, in the company of the

*Following the practice of Richard Dyer in his BFI monograph *Stars*, I shall refer to actors in films either by their name or by that of the character they play, whichever seems the more important in the particular context; where (as here) both are crucial, both will be given in the form shown, the actor/actress's name coming first.

homesteaders' adopted son Martin, begins as a mission of bloody revenge, whose object is to trace and kill Debbie because she is polluted by association with her Indian captors. When she is finally found, however, Ethan lifts her up in his arms and says: 'Let's go home, Debbie.'

The goal of revenge has changed (as we may well have suspected all along) into one of restitution and harmony. At the end, Wayne returns her to her surviving relations before plodding out of the frame towards an unknown destination. The colonists' harmony, disrupted at the beginning, is reasserted; the contrast between the 'garden' and the 'wilderness' often seen as central to the whole Western genre[2] is resolved in favour of the former (literally in the rebuilding of the homestead and figuratively in the transformation of Ethan Edwards's heart); and the only loser, apart of course from the massacred parents, is Edwards himself, whose departure signals that there is no place for him in the order of things whose balance he has helped to restore.

Certain fairly obvious points, common in varying degrees to the whole genre, immediately suggest themselves. Right and civilization are unequivocally on the side of the white man, and the Indians are portrayed as gratuitously bloodthirsty barbarians. The white settlers' society is resourceful enough to overcome such little local difficulties as the butchery of many of its members and assert a resilient sense of community (prefigured in the singing of 'We Shall Gather at the River' in the early funeral sequence). John Wayne is, as ever, a big man, not only physically but in human terms; it is his bigness that enables him to survive the gruelling trek and that makes his final softening towards Debbie at once so moving and so inevitable. A sense of organic community, a racism so unquestioned that it required neither justification nor articulation, a constant stress on what it means to be a 'real man' – here are features characteristic of much in American society, and nowhere in its culture more pronounced than in the Western. But how can the more complex type of analysis I have been propounding help us to get to grips with the specific inflection these themes receive in *The Searchers*?

In the first instance, the gap between the time of the film's first release and the present day is a period in which the image of John Wayne has undergone multiple transformation. His early

Introduction

quality of *macho* abrasiveness mellowed at the turn of the forties, in his work for Howard Hawks (*Red River*) and Ford (*Rio Grande*); even so, the suddenness of his transformation from avenger to fond uncle might still have surprised a 1956 audience, for whom his ferocity as Tom Dunson in *Red River* would probably have been a more enduring memory than his softening at the end of that film. An informed modern audience would be likely to read this aspect of *The Searchers* in a more complex and cynical way. Wayne's unflinching involvement with the most hawkish brand of American right-wing politics, culminating in his full-blooded support for the Vietnam war, has irrevocably changed the way in which modern filmgoers perceive him. Ethan Edwards's cradling of Debbie is now likely to be read in much the same way as Wayne's kindliness towards the small Vietnamese boy in *The Green Berets* – as a rather cynical camouflage for politics whose underlying brutal racism is in fact fundamentally responsible for the strife and conflict that pervades both films.

But this is by no means the full story. If it were, it would be extremely difficult to account for the esteem *The Searchers* has consistently enjoyed among audiences and critics (such as the French *Cahiers du Cinéma* group) to whom the macho and racist attitudes we have detected are anathema. One important reason for this is Ford's status as an *auteur*.* However discredited this theory may be today, with its stress upon autonomous individual creativity in a medium whose costly complexity makes such an approach only partially valid, it certainly helped to bring the themes that consistently recur in certain directors' work into prominence; examples in *The Searchers* would be the periodic reassertion of community through ritual and the use of 'real' physical landscape (in a part of the West Ford knew and loved) to comment on the characters' development and attitudes. Because these themes have come to the fore thanks to auteurial criticism, the possible ambiguities in their representation of the film's society are more clearly visible. Questions that could be posed about the film in this light would be such as:

– How far is the ritual reassertion of community 'just' a Fordian artistic device, and how far a paradoxical revelation of how

---

* For a working definition of this and other technical or specialist terms used in the text, see the Glossary.

precarious and insecure the very idea of community is/was in American society?

– Is Wayne/Edwards's departure from the restored world of domestic happiness at the end of the film 'just' to be seen as a rhyme to his arrival at the beginning, thereby bringing the film's epic action full circle? Or does it rather derive its specifically elegiac quality from the ultimate absurdity of a Western world in which John Wayne, of all people, can find no place?*

The formal and aesthetic questions raised by an auteurist analysis, in other words, debouch directly onto others which show how the film constantly undercuts the assumptions on which it might be thought to depend. The relationship between the pursuing Edwards and the object of his pursuit, the Comanche chief Scar, poses another series of questions, most strikingly in the scene where Edwards scalps his hated rival after his death. The moral issue raised – that Edwards's wrath has caused him to sink to the same level as his enemies – is clear enough, and can be read side by side with the fundamentalist Evangelism that recurs in the burial- and wedding-sequences. But if we pursue this further we come up against such questions as:

– It is revenge, and the desire quite literally to get his own back on/from Scar, that motivates Edwards's pursuit. If this causes him to become in one very obvious way like Scar, does this not suggest a possible explanation for the otherwise unmotivated Indian brutality which functions as an unquestioned assumption of the film?

– Edwards begins the film (as he ends it) as an outsider, having apparently continued to fight a rearguard action on behalf of the defeated Confederates after the Civil War has finished. Are he and Scar not therefore in some sense mirror-images of each other, as outsiders to the ordered world that *both* in some sense disrupt?

– Edwards displays throughout the film a comprehensive familiarity with Comanche language (he can act as interpreter between Martin and his Indian would-be fiancée 'Wild Goose') and custom (as when he shoots a warrior's eyes out after his death so that he cannot find his way into the Comanche Promised Land). While this is understandable in one who is presum-

---

*This second interpretation again rests upon a knowledge of specific historical and social factors – in this case, that the late fifties were to mark the 'beginning of the end' of the Western as a consistently viable cinematic genre.

Introduction

ably a seasoned Comanche-hunter, does its frequent recurrence not suggest that he is in some way an 'honorary Comanche'? And is this not strengthened by his own inability to find his way into (or place in) the 'Promised Land' of the restored 'garden' at the end?

I do not want to 'reduce' the complex world of *The Searchers* to a series of examination-type questions (to anticipate criticism of this type of approach from those who have no qualms about such modes of analysis applied to literature). Rather, I want to show how careful consideration both of our own time and of that of a film's making; of the social and historical implications behind what are often presented as purely formal strategies; of the manner in which genre assumptions quite unacceptable to the sensitive liberal hearts of many viewers and critics can actually undercut, or even destroy, themselves within specific individual films; and of the multiple ambiguities that cluster round a genre such as the Western (or for that matter the Japanese samurai film), and an actor such as John Wayne (or Mifune Toshiro), can all help us to a clearer, because richer and more complex, perception of how national cultures and identities appear in the cinema.

A final brief paragraph on the language problem. Films are generally referred to by the title under which they have been commercially available in Great Britain – thus, most French films bear their original titles, while the Japanese ones vary. There is no good reason why *Chikamatsu Monogatari* and *The Story of the Last Chrysanthemums* should have been released in Britain with titles in different languages, but the fact remains that those are the forms in which British audiences are most likely to have encountered them. Where there is doubt even here I have bracketed original and English title (as with *Ohayo/Good Morning*). Japanese names have been spelt in English transliteration, but using the Japanese word-order, with surname before 'Christian' name (thus, Kurosawa Akira not Akira Kurosawa). Finally, films and books/articles alike are referred to in text and footnotes only by their name. Fuller details of date, author/director, publishers, etc. will be found listed alphabetically in the bibliography and filmography.

Part One

The Organization – American Society in and through its Cinema

A study such as this has, inevitably, to begin with the American industry (referred to, for convenience, as 'Hollywood' in this work, though it should not be forgotten that many films, such as the New York location thrillers of the late forties and early fifties, are important largely because they were *not* shot in California). Cinema means Hollywood for most cinemagoers in the West, and not for reasons of productivity alone; the volume of films made by Hollywood would have been impossible, or of no consequence, without the international publicity and distributional organization that backed them up, transforming what began life as a technical divertissement into the most pervasive and successful agent of cultural imperialism there has ever been. The star-system, the studio-system, the changing-yet-codified system of film genre – all form a largely unquestioned part of what we understand by 'the cinema', through their export and reproduction in other industries as much as through their virtual omnipresence in the American commercial film.

Particularly important in this connection is the way in which such Hollywood-centred concepts have dominated most writing and broadcasting on the subject of film. To talk of 'Francis Coppola's *Apocalypse Now*' as though such a film could unproblematically be assigned to one creative source; to think on the other hand of 'Fred Astaire's *Swing Time*' despite the fact that the directorial credit for that film belongs to George Stevens; to programme repertory-cinema or television seasons of horror movies, or 'classic' Westerns, or *films noirs*, in the awareness that your audience will know what to expect, and to a large extent select itself on the basis of the genre on offer (witness the existence of sub-groups such as 'Western addicts' or 'Hammer fans') – these assumptions are the kind of conditioned reflex by which the values and ideologies latent, and sometimes patent, in Hollywood cinema are perpetuated. Not that there is anything wrong in trying to assign praise or blame for a film to one individual or group of individuals in particular, or in dealing with a number of different-but-similar films under this or that genre heading. I shall constantly be doing both throughout this study. What is important is to be aware that these categories and modes of thought are not 'innocent', but ideologically produced and inflected, and that images of American society are produced, not merely by the individual films themselves, but by the structures of the film industry and the society as a whole and by those, closely allied, of the ways in which we tend to think, talk, and write about films and about those who produce, direct, appear in, and otherwise 'make' them.

There is a common denominator here, and it turns out to be, unsurprisingly, one of the dominant notes constantly sounded by Americans writing in praise of the society in which they live. Culturally it can be summarized as the idea of unity through diversity (the 'melting-pot'); politically, as that of a prized democratic freedom of choice from among freely competing groups (the 'political marketplace'); economically, as that of discipline and organization through untrammelled individual competition (with government as 'umpire' rather than participant). What these three variations of one stance have in common is belief that efficient organization and *laissez-faire* rivalry are complementaries not opposites, that cultural, political, and economic groups left unhindered to compete among themselves for resources, votes, or markets (or, alternatively, to amalgamate or integrate on a voluntary basis) produce the highest and most widely satisfactory degree of organization that a society could wish for without tyranny.

This ideology finds its most extreme expression in the glutinous blockbusting novels and bloated 'philosophical' works of Ayn Rand, one of whose heroes, lone rebel against a society of tyrannized conformists, traces the dollar-sign in the air as a proud badge of human freedom.* Rand's importance in forming

*Masochistic curiosity can be satisfied by reference to *The Fountainhead*.

'Middle-American' opinion should not be underestimated, and it is worthwhile pointing out her enthusiastically collaborative role in the McCarthy investigations in the fifties, which ruined many Hollywood careers by indiscriminate and paranoid use of the two adjectives 'Communist' and 'un-American'. But a far more sophisticated and influential articulation of similar ideas can be found in the work of Milton Friedman, from whose *Capitalism and Freedom* the following passage is taken:

> So long as effective freedom of exchange is maintained, the central feature of the market organisation of economic activity is that it prevents one person from interfering with another in respect of most of his activities. The consumer is protected from coercion by the seller because of the presence of other sellers with whom he can deal. The seller is protected from coercion by the consumer because of other consumers to whom he can sell. The employee is protected from coercion by the employer because of other employers for whom he can work, and so on. And the market does this impersonally and without centralised authority . . . The characteristic feature of action through political channels is that it tends to require or enforce substantial conformity. The great advantage of the market, on the other hand, is that it permits wide diversity. It is, in political terms, a system of proportional representation. Each man can vote, as it were, for the colour of tie he wants and get it; he does not have to see what colour the majority wants and then, if he is in the minority, submit.[1]

The direct equation of economic (and, by implication, cultural) with political freedom that is manifest in this passage is a leitmotif of Friedman's thought – which means of a tendency that, in less carefully conceptualized terms, has been dominant in American society for much of this century (witness the widespread branding as 'Reds' of those who advocate even the most moderate forms of socialized health care). In the introduction to the same work, Friedman defines the system he advocates as 'the *organization* of the bulk of economic activity through private enterprise operating in a free market . . . a system of economic freedom and a *necessary condition* for political freedom' (my italics).[2] Purveyors of (in our case) cultural wares, in other words, can only be left free to organize themselves through the absence of externally imposed organization and the checks and balances of producer and consumer freedom. Any attempt to do otherwise will lead to the simultaneous and inexorable erosion of the material base of their industry and of their right to say, print, or film whatever and however they choose.

The relevance of this to the Hollywood phenomenon goes beyond such obvious areas as the McCarthyite purges. For, despite multiple interference in the area of censorship (of which the Hays Code of 1930, which made overt reference to sexuality virtually impossible, and the taboo on explicit social criticism after McCarthy are the most notorious instances). Hollywood has throughout its existence functioned very much in accordance with the free-market model propounded by Friedman. The most significant apparent exception to this is the various attempts made to break the stranglehold of the large production and distribution concerns, notably the anti-trust laws of 1915 and after the Second World War – but the exception is only apparent, for the action taken by the courts was that of an umpire, responding to appeals from smaller and independent companies that their larger rivals were abusing the rules of free competition.

But our concern is not only with the way in which the directly economic functionings of Hollywood exemplify the capitalist ethos of organization-through-competition. The very forms with which Hollywood is synonymous are also based upon it. What are the star- and studio-systems but individuals and corporations vying for dominance at the box-office and in audience consciousness, and thereby organizing the field so that MGM came to be equated with colourful musical extravaganza, or Humphrey Bogart and James Cagney to confer a clear 'brand-image' on their films and thus help to structure the perception of audiences?

The rather more ambiguous domain of genre is also important here. Most people operate the concept almost unconsciously (everybody recognizes a Western when they see one), and for a long time it was the dominant mode of thinking and writing about American cinema. Recent developments in the study of cinematic ideology, focusing more on detailed analysis of individual film texts, have tended to relegate it to a subordinate position from which it is only just beginning to emerge. Since our perception of American society through its cinema will be greatly influenced by our understanding of its different genres, it is worthwhile asking what exactly 'genres' are and how they show the influence of the capitalist ethos.

Stephen Neale, in his BFI monograph on *Genre*, sees genres as 'systems of orientations, expectations and conventions that circulate between industry, text and subject'[3] – that is, between the producers whose goal is to make a profit, the individual film which will only attain that goal if it is in some way (usually by genre among others) recognizable yet also different from its predecessors, and the members of the audience who will need to be able to identify themselves in some way in or with the film if it is to mean anything to them. Genre, in other words, binds these three parts of the film-viewing process together, confirming that the film before the audience conforms to a familiar type ('another spine-chiller from master of suspense Alfred Hitchcock' – an example in which the director's name also serves to reinforce indentification), yet is in some way a development of or addition to it.

Antony Easthope, in the winter 1979/80 issue of the magazine *Screen Education*, sees genres as forms of consolidation – for the industry, to ensure a profit from the films it turns out, but also for the audiences:

> On one side genre is an 'optimal exploitation of the production apparatus', an economy of manufacture according to stereotypes imposing their own work discipline and corresponding perhaps to the assembly line in other industrial production. On the other side genre acts to maintain hold on the consumer through a standardised product and consistent reproduction of a successful product.[4]

What is clear in both Neale's and Easthope's formulations is the importance and desirability, for producers and consumers alike, of the notion of organization, and the vital role played in this respect by genre. What exactly a genre *is* it is almost impossible to say, since each addition to the field is bound to change it in some way; Jean-Louis Leutrat speaks of it as 'a memorial metatext ... one single continuous text',[5] that is, a crystallization of audience expectations (almost a collective unconscious), that unites different individual texts into one.

How can it be said that the organization of the industry by genres is 'competitive'? Clearly the various genres are not trying to drive one another out of business. Nor, however, is this necessarily what capitalist firms try to do; they may well find it in their best interest to specialize each in a particular sector of the market, which will therefore become subdivided into a number of only nominally competing enterprises. Something very like this happened in Hollywood in the thirties and forties, where horror pictures were the province of Universal, cut-price Westerns that of Monogram, and so forth. But, whether or not the division of the field by genres was accompanied by a corresponding division by companies, the principle of organization through the market has remained valid for most of Hollywood's history. Not only individual actors, but directors and theatres came to specialize in particular genres (for example, the identification of John Wayne or Bud Boetticher with the Western, the 'brand-image' of different cinemas in all major cities specializing in art-cinema revivals, or late-night horror, or – unfortunately at the moment the commonest example – sex-ploitation). Audience tastes and expectations have stemmed largely from identification with what the market has to offer (which gives the lie to the Friedmanite notion of 'freedom' – the market exists to form tastes as much as to satisfy them). Competence to 'read' a film is dependent upon familiarity with the conventions and iconography of genre; these in turn, by the degree of stylization (sometimes even extending to interchangeability of sets and footage) they make possible, have helped to keep production costs down and thus maximize profit. The low-budget black-and-white décor and shooting of *film noir*, and the apparently 'simplistic' but in fact audaciously stylized use of colour in Douglas Sirk's films for Universal, exemplify this tendency.

I selected these two examples because they also serve to illustrate another important feature of the Hollywood 'marketplace' – the aspect of 'secondary commercialization' referred to in the introductory chapter. *Films noirs* were usually B-features, acting as support to more lavish and better-publicized films. Many of them now enjoy considerable reputations on the 'art-house' circuit and its close relation, the late-night television programme. Sirk's melodramas, likewise, originally envisaged (by the studio at least) as potboiling 'women's pictures', are now, because of their visual audacity and penetrating analysis of family problems, not only art-house successes, but fundamental texts for

Marxist and feminist criticism. This has two main consequences for our analysis. First, that the workings of the market are not quite so innocent as its advocates would have them appear. These films have realized their production costs many times over in two distinct and separate contexts, and on each occasion precisely because of the enforced economy which went into their making. It is the grainy black-and-white shooting of *film noir*, and the lack of realism in Sirk's studio sets – both largely economically defined qualities – that have attracted fresh audiences to these films, so that the necessary parsimony in their making has turned out to be a double investment. On one level at least, they can be read as an eloquent testimony to the 'American' values of shrewdness and economy – a reading of course belied by most of their other aspects.

The second consequence is that it can be very misleading to speak of a particular film as though it always and everywhere produced the same meanings to be read in the same way. Much analysis of film tends to treat individual texts as immutable repositories of meaning, a methodological lapse which often hardly matters (and which is to some extent inevitable if films are to be meaningfully written about at all), but which – particularly with Hollywood films that have undergone secondary commercialization – can be positively dangerous. A New York audience of 1943, two years after the USA's entry into the Second World War, watching *Casablanca* in the year of its release, would have been likely to foreground the film's numerous references to cynical neutrality versus commitment (notably in the persona of Bogart/Rick, the former fighter for good causes who now 'sticks his neck out for nobody'), and to react accordingly to the atmosphere of cosmopolitan self-interest and amoralism – a whole world of politico-diplomatic neutrality – that characterizes Rick's Café. Such factors are probably a long way from the minds of members of university film-societies who nowadays regularly provide the film with capacity audiences, and whose admiration for the tough-yet-tender anarchistic persona of 'Bogey' is filtered through nearly forty years of camp cultivation of and writing about the movie, so that the mere fact of going to laugh and cry at *Casablanca* for the *n*th time has become a kind of cultural statement. Neither of these readings is 'right' or 'wrong'; but, if we are to be truly aware of how American society is depicted in its cinema, the first, historically distant as it may now seem, needs to be borne in mind as much as the other.

So far we have concentrated on factors outside the individual film-texts themselves. But we have also to bear in mind that American society – its history, its politics, its economy, and its culture – is constantly present within these texts, as well as impinging upon them from outside. This is obvious enough when we look at the importance of Prohibition in the gangster movie, or of white racism in the Western, or of urban and political paranoia in the work of such seventies directors as Altman and Pakula. But there is a deeper sense in which the statement is true, connected with the ethos of organization-through-competition that is central within (as well as to) the main Hollywood genres. This ethos is perhaps plainest in the gangster movie, where villains and the forces of law-and-order compete for the economic and social domination of society, and plainest of all in some of that genre's later manifestations, where gangsters and police are constructed as apparently identical organizations, or the Mafia revealed as but a large-scale structural variant of that fundamental mode of American social organization, the family (*The Godfather*). But it appears elsewhere too. What else are the archetypal situations of the Western – the skyline confrontations between lone hero and lone villain, the deadlock between frightened townspeople and marauding rustlers resolved only by the arrival of the charismatic individual hero – but variations on this theme, made more uncompromising still by the absence or powerlessness of any governmental umpire? What is the terrifying – yet also perversely exhilarating – world of *film noir* but the other side of the universe of unbridled competition, sexual Darwinism here coming to reinforce its social counterpart? The musical may appear to be an exception, but it is interesting to note how many musicals have plots concerned with competition to succeed (*Singin' in the Rain*) or merely to survive (the *Gold Diggers* series) in the cut-throat world of entertainment (presented more or less explicitly as a paradigm of American society). And competition (for the girl's affection, the favour of producers, or simply the attention of the audience at any particular moment) provides the motivation for a good many of the set-piece musical numbers.

Colin McArthur has commented interestingly on two important Hollywood genres in this connection: 'It could be said that they represent America talking to itself about, in the case of the

Western, its agrarian past, and in the case of the gangster film/ thriller, its urban technological present.'[6] What is important here is not only the relationship between history/society and film genre, but also the notion of 'America talking to itself'. America talks to itself about the capitalist system of free enterprise, in the works of a Friedman or a Rand, in such a way as to suggest that that system will produce the best of all possible worlds, and that its conspicuous failures are the result, variously, of excessive state interference with the market machinery or of human weakness, fallibility, or corruption. But a society talks to itself differently in overtly conceptual works and in works of art (however defined). A conceptual work – an economic or political text – normally seeks to *state*; a work of art can normally only *show*, which despite appearances is the very reverse of a weakness. This is because the contradictions within a society, often explained away or 'invisibly mended' in more expressly theoretical discourses, are in a work of art often left gaping and unresolved – and, even if a resolution is attempted, the conventions of fictional discourse deprive it of the legitimacy, the intellectual and pedagogical authority, which Friedman (or Adam Smith, or for that matter Karl Marx) derive for their conclusions because of the very type of text they are writing.

Literary criticism has often focused on the discrepancies between the known political opinions of a writer and the ways in which his texts contradict these.* For an observer of Hollywood, this terrain is a rich one indeed. The competitive ethos on which Hollywood was founded, and the various types of censorship and exclusion at work in the industry, always tended to exclude or marginalize those most vocally critical of American society (Chaplin's departure for Switzerland, and subsequent fraught relations with the Hollywood establishment, is the most striking example).

The overwhelming ideological tendency in the industry has been to the right. Its genres, we have seen, are largely predicated upon the virtues of competition and the survival of the fittest. Yet all sorts of uncomfortable questions present themselves for anybody seeking to show that the American commercial film is unequivocally committed to a defence of the

---

* Georg Lukacs's work on Balzac (in *Studies in European Realism* and *The Historical Novel*) is the best-known example, but Dostoevsky and Joseph Conrad have both been the focus of similar attention.

status quo. Why, for example, in the supposedly macho Western, should the charismatic individual leader-figure lapse into near-psychopathy and need to be taught a sharp lesson (John Wayne/Tom Dunston in *Red River*), or stand out as a salutary contrast to a cowardly and self-interested mass (Gary Cooper/Kane in *High Noon*)? Why, in the gangster movie, should the happily Americanized European family be capable of harbouring a matchlessly efficient crime machine (*The Godfather*). or the supposed salvation of the land of the free from Communism be the work of those whom that land has despised and rejected (*Pick-Up on South Street*)? Why should devotion to one's mother – perhaps the archetype of all American civic virtue – be directly indicated as the source of psychopathy (*White Heat*, *Psycho*)? Why should an apparently happy line of chorus-girls singing 'We're in the Money' be shortly afterwards thrown out of work – and, by implication, close to prostitution – by the Depression (*Gold Diggers of 1933*)? Competition's gaping holes – its inhumanities, immoralities, inbuilt systems of discrimination, and (ultimate insult) inefficiencies – remain disturbingly open in its most characteristic commercial/artistic products as they do not in more conceptually sophisticated types of discourse. It is hardly likely to be a coincidence that the Hollywood films most widely judged interesting are those in which the various bankruptcies of the Great Society most strikingly reproduce themselves.

The Silent Era

Right from the first American film screening in 1896, the value of the medium in constructing an image of the country, especially for its more disadvantaged groups, was plain. Immigrants with little or no English, hitherto shut out from most American entertainment-forms, made up a considerable proportion of early film audiences. As Lewis Jacobs says in *The Rise of the American Film:* ' . . . more vividly than any other single agency they (films) revealed the social topography of America to the immigrant, to the poor, and to the country folk.'[1]

Common themes in short pictures at the beginning of the century seem to have included the prevalence of social injustice, the courage and even heroism of the 'common man', and the demoralizing effects of poverty – dealt with, however, from a moralizing rather than a socially analytical point of view. This may well have been due to the narrative limitations of the short film, for in the feature films of D. W. Griffith (the earliest American features at all widely available today), the contradiction between an ethical and a socio-historical perspective is marked, perhaps the result of the encounter of Griffith's broad historical sensibility with the individualized forms of the short film.

The Birth of a Nation (1915) recounts the main events of the Civil War through their impact on two families (one Northern, one Southern), and in its second part melodramatically exaggerates the decline of Southern 'emancipation' into ruthless exploitation by the carpetbagging industrialists and bankers of the North. This section, culminating in a cavalry-style ride to the rescue by the Ku-Klux-Klan in full regalia, has tended to dominate modern critical reaction to the film by its racist offensiveness, acutely felt by coloured audiences when the film was released (the National Association for the Advancement of Coloured People protested vigorously). In fact there is an ambiguity here that is to be fundamental to much subsequent American cinema (particularly in the Depression years), that of the 'populist' approach, which castigates the injustice done to the 'ordinary people' by their social superiors, concerned only with power and money, but unproblematically aligns itself with the ordinary people's supposed point of view, including all too often their prejudices.

The Birth of a Nation criticizes the market organization of embryonic American industrial society (the victory of the North had been that of industrialization over quasi-feudal Luddism), but adopts what may well have been the only perspective available to Griffith at the time, one impregnated with racial condescension. I say 'condescension' rather than 'prejudice' advisedly, for the 'bad' Negroes in the film are shown as putty in the hands of their carpetbagging manipulators, and the portrayal of the 'good' ones who cleave to their masters at all costs is hardly less offensive. The Negro, loyal servant or lascivious monster, is what the white man chooses to make of him. Griffith in fact constructs a universe within which three types of social organization struggle for dominance: the 'pre-civilization' of the Negro, relating to his 'superiors' only on the plane of submission or instinctive defiance; the idyllic but menaced stability of the Old South; and the thrustful industrialization of the North, bringing in its wake corruption, despair, and the destruction of its competitors. The curses of market rivalry are more apparent here than its blessings.

Erich von Stroheim's attack on the cult of the dollar in *Greed*, launched ten years later in 1925, is much more overt. There had been many changes in the interim, notably the feverish expansion of the economy and the definitive emergence of California as the centre of the cinematic industry. Considering the insuperable financial problems which Stroheim was suffering by this stage in his career, it is not implausible to see in the ruthless California-centred avidity and social Darwinism of its main characters a bitter comment on the way in which Hollywood stifled the individual talents it purported to encourage. The misfortunes of its protagonist, McTeague, start when he is prevented from practising as a dentist because he has no professional qualifications. The unqualified (and often dangerous) medical practitioner is a common early Hollywood figure (for example, the Western quack-doctor, or W. C. Fields's manic tooth-puller in *The Dentist*). It is gratifying here to see him restrained by society, horrific though the consequences are. At the end of the film, McTeague (who has murdered his pathologically avaricious wife, stolen her hoard, and fled into the desert), fights to the death with his pursuer and rival in the arid Death Valley – the stake a bag of gold coins, the certain outcome fatal for both since they are miles from the nearest water, the futility sentimentally symbolized in McTeague's kissing and release of his long-suffering pet canary.

Explicit 'social criticism' has long since been overlaid by Stroheim's iconographic sadism, but this too has its ideological dimension. The close-up was the only means available to silent film, given the cumbrousness of intertitles, to convey the emotional states of its characters; at the same time, it inevitably did this by divorcing them from their context and thus seeming to provide the spectator with a direct and privileged vantage-point upon their interiority.* Films such as *Greed*, or Dreyer's *The*

---

*This accounts for its prominent place in the writings of humanist-realist critics such as Bela Balazs.

Passion of Joan of Arc, turn this to great effect, largely because the social context in both is abnormally exiguous (the Dreyer film takes place within the sombre surroundings of a medieval trial, and towards the end of *Greed* the social focus narrows down – in part due to ruthless studio editing – to the interaction of a few individuals in increasingly stripped and arid settings). But it remains true that the thrust of the close-up in general is towards isolated consideration of individuals, and that until the coming of dialogue it was very difficult for the cinema to articulate individual emotional states with social and historical movements. Griffith's often-mentioned 'emotional naïvety' contrasts with his epic sense of history to provide confirmation of this.

The area of silent cinema where the relationship between the individual and society is for a contemporary audience most eloquently articulated is, perhaps surprisingly, the comedy film. Mack Sennett's early Keystone Cops shorts – immensely popular with immigrant audiences – depict the city as a terrifying world of people and objects out of control, their slapstick deriving its effect precisely because it acts as a resolution of the tensions this induces.

The mechanization of American society is even more alarming, because seemingly directed against one hapless individual, in the films of Buster Keaton. One gag in his Civil War film, *The General*, will serve as an illustration of this. Near the beginning of the film Buster has been rejected as a volunteer for the Confederate Army because (though nobody troubles to tell him this) he is felt to be more valuable in his civilian capacity as a railroad engineer. Marion Mack, his lady-love, spurns him as a result, and Buster sits ruefully on the cross-bar of the locomotive after which the film is named. He fails to notice that the engine is beginning to move, and still lost in dismay is carried into a tunnel . . . to emerge, the next title tells us, 'a year later, in a Union encampment just North of Chattanooga'. The gag is funny yet tender enough to be one of Keaton's most frequently retailed, but it has implications for the whole problem of the isolated individual in an increasingly mechanized society.* Not only is Buster rejected (in perfunctory and inhumane fashion) by the machinery of the Southern war effort; that social rejection leads directly to a sexual one, and that in turn to his being – literally –

---

*I refer here, of course, to the society of the time (1926) at which the film was made and first screened, not to that in which it ostensibly takes place.

taken up in the machinery of the railroad (his peacetime occupation), and deposited in the heart of the action at the beginning of the next reel. The choreography – as always with Keaton – is masterly, and the deadpan expression serves to concentrate our attention upon the mechanical organization that surrounds him. It is the latter that here, and in many other places in his work, provides the focus for audience identification.

There is undoubtedly a connection between the very high proportion of comedy shorts and features made during the silent period and the prominence in them of the 'man-versus-machine' theme. Audiences (drawn in this 'nickelodeon' period overwhelmingly from the less well-off classes) were clearly attracted to evocations, especially comic ones, of an urban society which struck many of them as harsh and brutalizing, whatever its merits in terms of profitability might be. This perception, most subtly and masterfully organized in the films of Keaton, comes close to urban paranoia in the work of Harold Lloyd, climbing skyscrapers or clinging frantically to hands on a clock-face, and is given a more consciously endearing inflection in the important persona of Charlie Chaplin.

Lewis Jacobs wrote of Chaplin in 1939 that 'his frustrations are mankind's; his successes, universal triumphs'[8] – an ironically-timed observation in view of the conflict that was about to rip apart any consensus notion of what 'mankind' might be, but one that chimes with a widely accepted view of the 'Little Man'. While the characters he adopts – failed goldprospector, maltreated immigrant, reified factory-worker, or best-known of all that ultimate social non-person the tramp – offer a wide range of possibilities for audience identification, there is nothing specifically American about him, or even specifically working-class. The grace and elegance of his movements, the manicured charm of his gestures and features, and his mixture of surreptitious defiance of his oppressors and timorous deference towards them are all more evocative of a middle-class attitude (upper-middle-class as far as the walk and gestures go, petit-bourgeois in the attitude that underlies them). The resulting mixture of working-class situations and middle-class behaviour within them may well account for the double tenacity and durability of Chaplin's appeal – across classes and down generations. A British-born music-hall comedian who spent the latter part of his life in Switzerland, Chaplin has little overtly American about him, but might be thought of as (in class terms) an attempt at a one-man 'melting-pot'.

Women, in the films so far discussed, play at best a secondary role, prizes to reward the successful hero or elude the comic dupe. But autonomous female sexuality made its appearance in the American cinema remarkably early, helped by the First World War, which led to the breakdown of many old ethical restrictions at the time when moviegoing was reaching its peak. Theda Bara, the original 'vamp', was extremely popular during the war years, and her work inaugurated a strategy that Hollywood was to use more and more as censorship tightened its grip after the coming of sound. Her films were often provided with ostensibly 'moral' narrative resolutions, which were unlikely to fool most audiences; and it is in the conflict between desire for vicarious erotic adventure and contemplation of female sexual attractiveness on the one hand, and reassurance of and within the established social order on the other, that such contradictory resolutions, laughable to present-day audiences, are founded. An extrovert, self-made hero such as Douglas Fairbanks might have won as his prize – on the screen and in real life – the demurely vulnerable Mary Pickford, and the average male filmgoer probably wished for nothing better than to marry such a wife. But this did not stop him from flocking to enjoy fantasies of illicit dalliance with a Theda Bara whose flesh-and-blood presence would have terrified him. The contradictory resolution of so many sexually charged Hollywood pictures owed as much to the internal 'censorship' of the libidinous yet ultimately conventional American male as it did to more external forms. The movies early became a means for American society to reconcile in fantasy the sexual contradictions its reality had difficulty in accommodating.

A contrast parallel to the sexual one just described, and equally marked in films towards the end of the silent era, was that between social realism and escapist fantasy. King Vidor in *The Big Parade* and *The Crowd* showed a supposedly ordinary American male (the type often described as a 'Good Joe') undergoing tribulations in war and peace respectively; but by this time, just before the transition to sound, the social-realist movie had tended to become the exception. Even *Greed*, which is hardly escapist, reveals in its stylized contortions the need for the camera to move away from ordinary life. Unbridled sexuality could be partially defused by being transplanted into exotic settings, distant in time or space (as in the films of Rudolph Valentino). As the time and resources available for leisure increased, so the movie industry concentrated more upon it. Even the allegedly wild West was tamed and civilized by a British butler in *Ruggles of Red Gap*. The society suggested through films of the twenties was one of lavish and slightly self-mocking indulgence.

Industrial and political issues, if Hollywood was to be believed, were all a matter of mediation and understanding. In countless films now forgotten the bitterly polarized extremes of management and labour were reconciled through the moderating agency of goodwill (often incarnated in a woman or a priest). This might not have been so offensively mendacious had the films not taken place against a background of political repression involving the shelving of the Bill of Rights for much of the twenties, anti-'Bolshevik' purges little less ferocious, if less systematic, than McCarthy's thirty years later, and government pressure on the industry to portray 'Reds' in a hostile light. These denials of democracy, at a time of largely artificial stimulation of the economy, were more effectively occluded in the films of the period than the denial of sexuality, which as we have seen could not entirely erase the fond memory of its indulgence. A society of benign industrial cooperation and playful erotic escapism, in which real pain or tension could always be soothed away by the appearance of a suitable mediating figure – such was the vision that films of the end of the twenties attempted to construct. The pressure upon them to do so, needless to say, derived from sources rather more interventionist than the play of market forces – notably the government of President Harding (1920–23), isolationist, protectionist, and paranoid about 'Reds' in the wake of the Russian Revolution.

The Early Sound Period and the Films of the Thirties

The Russian Revolution, oddly enough, was also a contributory factor in the rapid acceptance of sound by the American industry after *The Jazz Singer* of 1927. The quality and inventiveness of the Soviet films that reached the United States made the industry realize the competition it was facing internationally, and the benefits of sound in other respects were too obvious for resistance to it to be long-lived.

How did sound affect the depiction of American society in the cinema? For spectators of today, it has almost certainly foreshortened their perception of it; so swiftly did silent films become obsolete as objects of consumption as well as production that very few of the films mentioned hitherto (other than comedies, which have remained popular thanks to television) will have been seen by any but specialist audiences.

In terms of the structure of the industry, sound favoured the development of certain types of film (it is difficult to talk about 'genres' in the silent era, if only because relatively less is known about its films), and was the virtual kiss of death to others. The possibilities of the animated cartoon, to become for Disney the most all-pervasive agent of Hollywood imperialism, began to be explored more fully. Soulful melodramas in romantic settings became far more difficult to take seriously once words started issuing from the actors' mouths, yet the social factors that produced a guaranteed audience for escapism were still at work. Where did this audience go, and how did sound change the kind of films it watched?

Part of the answer is provided by the extraordinary popularity of the musical. This was so obvious an area for the new sound cinema that the market quickly became saturated; by the early 1930s it was no longer enough to take two established stars, place them in an exotic studio setting, and set the cameras turning to be assured of a hit. The genre had begun as a fusion of Europe and America, the operetta and the Broadway musical-comedy revue; the former was responsible for the continued note of escapism (exemplified in Jeanette MacDonald's singing 'Beyond the Blue Horizon' from a railway-carriage window for Ernst Lubitsch in *Monte Carlo*). More interesting to a contemporary audience, however, is the Broadway influence, which often extends to the plots and settings of the films. The 'backstage story' was useful to producers and directors in a number of ways. It provided a ready-made pretext to film musical numbers, as part of the show whose presentation the film was about. It played upon audience fascination with the glamour of theatre life, often via a plot involving injury or other problems with the show's star and her triumphant replacement by an unknown understudy. And the supposed equality of opportunity whereby Fate always created the chance for a gifted unknown to show her talent nourished for millions of female cinemagoers the illusion that stardom was just around the corner, and that American society really would allow its most gifted members to rise to the top.

Exemplary in this respect are two Warner Brothers musicals of 1933, *Gold Diggers of 1933* and *42nd Street*. MGM was to become the leading musical studio, and Warner Brothers, under the strongly pro-Roosevelt aegis of their president Jack Warner, were in the thirties specializing in dramas with a 'social conscience'; this element is strongly marked in *Gold Diggers*, framed by two contrasting musical numbers that say much about the America of the Depression years. The overheating of the market had made a slump inevitable; when it came, in 1929, the cinema was to some extent protected, both because sound was still a novelty and because spending on entertainment pro rata to total income tends to be higher in depressed periods. But by 1933 Hollywood too was feeling the draught, and the *Gold Diggers* plot and numbers are an indication of this. At the beginning of the film, an excited group of showgirls is singing the euphoric 'We're in the Money'; this is immediately followed by the news that their show is to close, and that eviction from their lodgings will follow. The consequence of this for many of the girls' real-life counterparts was a choice between starvation and prostitution, hinted at but not made explicit in the film because of the constraints of contemporary censorship.

The 'gold diggers', needless to say, are saved, by the discovery of a wealthy young Bostonian whose only ambition, to the horror of his righteous family, is to become a composer of musicals. He writes and funds a show for the girls which concludes the film; that the rich really do have a heart is proved by the film/show's final number, '(Remember) My Forgotten Man', a plea for the post-World War I generation who found themselves on the scrapheap when the Depression came. It is interesting that contemporary audiences were scandalized by this finale, which was thought to be in the worst of taste. The film interestingly marries the escapist unreality of the early musicals (in the stroke of luck that enables the show to go on), with an awareness of social problems that deserves only praise considering the censorious climate of the time.

Censorship in the thirties was most apparent in the realm of sex. The Hays Production Code, drafted with the support of the Motion Picture Producers and Distributors of America Inc., and law within the industry by 1934, was instrumental in this, largely under pressure from the Legion of Decency (a Catholic Church 'front' organization). The Code pledged to oppose 'false, atheistic, and immoral doctrines', an objective whose anti-Communist thrust was clear, but which worked most effectively to suppress any explicit treatment of sexuality. If Clark Gable's reporter and Claudette Colbert's runaway heiress were carefully separated by an impromptu 'curtain' when circumstances forced them to share a bedroom in *It Happened One Night*; if the Negro was totally desexualized and reduced to a comic foil; if child-stars such as Baby LeRoy and Shirley Temple prospered, Hays was a large part of the reason why. The untenability of any rigid distinction between 'sexual' and 'political' censorship becomes clear when we consider the potentially shocking impact of miscegenation (class or racial), the determinedly idyllic presentation of childhood, and the suppression or narrative thwarting of any threat to the institution of marriage – all not just instances of puritanical hypocrisy, but conscious and deliberate attempts to block the representation of problematic areas of American society.

The sexual taboo was in fact easier to circumvent, especially for the more sophisticated metropolitan audience, than the more covert political one. The self-mocking glamorization of distance (in time or space) deployed by Marlene Dietrich in *Shanghai Express* or *The Scarlet Empress* still carried an unmistakable erotic charge; the choreography of Busby Berkeley (dance director of *42nd Street* and the whole of the *Gold Diggers* series) flaunted a plethora of pop-Freudian symbols and motifs – crane-shots of showgirls dancing to form vaginally evocative 'O's, the camera tracking under the chorus-line's open legs in the suggestively titled 'Young, Willing, and Healthy' number in *42nd Street*. Provided sexual transgression was not thoroughly explicit, the authorities – through naïvety or wilful, profit-motivated blindness? – seemed not to care.

Arcadia

Escape from the Depression, as the Berkeley musicals illustrate, was not incompatible with a partial acknowledgement of its realities. This appears most strikingly in films a long way removed from the domain of 'realism'. When Walt Disney's Three Little Pigs touched the hearts of a nation with 'Who's Afraid of the Big Bad Wolf?', it was because of their implied chirpy refusal to be downcast by the collapse of the economy; the Wolf huffed and puffed but failed to blow their house down, because they deployed the skills of energetic common sense against him. Disney's work, understandably neglected by many serious film historians, is in fact a rich repository of capitalist values.* The Three Little Pigs, with their vigorous voluntary cooperation against the monstrous threat of Depression (or Bolshevism?), must have gladdened the heart of Middle America.

King Kong's defiant rampage through New York in 1933 (the year of the Three Little Pigs as well as the Berkeley musicals) is another instance of how the cinema of fantasy can yield penetrating insights into a society's view of itself. The Depression years led to a reaction against big-city life, and the glorification of a simpler rural existence (in many ways comparable to recent trends towards dropping out and self-sufficiency). Roosevelt's 'New Deal' favoured this tendency by providing grants to set up small farms. The city was perceived as more and more threatening, and New York, in whose Wall Street the crash had started and whose skyscrapers bore eloquent witness to the feverish over-expansion of capitalism, embodied this perception. This is the background to, and the sub-text of, what can be regarded as the first 'disaster movie'. The giant ape Kong is no villain; he has been torn from the tropical setting in which he was venerated as a god and brought to New York, ostensibly to satisfy scientific curiosity but in fact to make money as a freak-show. His escape and orgy of destruction stem from his passionate devotion to Fay Wray, whom he caresses tenderly, and even makes to undress, while roaring defiance at bombers astride the Empire State Building.

Andrew Bergman, in his indispensable study of Depression cinema, *We're in the Money*, observes that films 'can present radically different models of behaviour with greater skill than they can possibly present anything like a "program" '[9] – a point reinforced by the French theoretician Christian Metz when he says that 'the film will always show things better, but the book

*This is fascinatingly explored in Armand Mattelart's *How to Read Donald Duck*.

will always say them better'.[10] It is necessary to go beyond this and say that what a film 'shows', or the models of behaviour it 'presents', can often lie well outside the explicit data of plot and character. Audiences of 1933 were very likely to be unaware – consciously at least – of the implications of *King Kong* (as they were not of, say, the Three Little Pigs). Modern audiences probably go to see the film because of its status as a 'Hollywood classic' and as a precursor of the contemporary disaster movie. In between these two modes of viewing the film – as contemporary entertainment and as an obligatory part of one's apprenticeship in nostalgia – a society's perception of itself may well slip away if not explicitly elucidated. We can say that *King Kong*'s durable commercial success is connected with its dislike of the world's financial capital; that the 'different model', in Bergman's phrase, it implicitly propounds is a world of almost bucolic innocence in which even giants are really 'noble savages' until goaded beyond endurance; and that, at a level which may well have escaped the film's makers completely, it also betrays profound disquiet at the destructive possibilities of sexual denial – the very denial shortly to be enforced upon Hollywood by the Hays Code and the Legion of Decency.

Innocence, in one form or another, is a central theme in the cinema of this period. It appears most obviously in the work of the arch-populist director, Frank Capra, who in *Mr Deeds Goes to Town* and *Mr Smith Goes to Washington* depicts the triumph of the ordinary man of goodwill over the corrupt machinations of finance and politics. Gary Cooper as Mr Deeds inherits a considerable fortune, through his innocent integrity wins the affection of the journalist (Jean Arthur) assigned to his story incognito, and uses his fortune to help struggling farmers, thwarting the attempts of incredulous relatives to have him declared insane. James Stewart as Mr Smith is nominated for election to the Senate because it is thought that his innocence will make him a passive, easily influenced member; he worships at all the major national monuments on his first trip to Washington, has no higher political ambition than to build a Boys' Camp, and displays extraordinary ingenuousness at every opportunity, yet it is this very quality which enables him to denounce intrigue and corruption from the floor of the Senate at the end.

At one level the populism of Capra's heroes can be read as mildly subversive, but this is only skin-deep. It is part of the market-capitalist ethos that free competition among individuals, in business or via the ballot-box, is the best possible method of denouncing manipulation and corruption, and that inequalities, even fortuitous ones, of wealth and power are necessary safeguards of freedom and reservoirs of individual influence over potentially tyrannical authority. In this respect, Deeds and Smith (whose names are significant – the man of action and the common man *par excellence*) exemplify that ethos very clearly, and their cultural values, such as Deeds's habit of playing the tuba for relaxation, draw upon a nostalgia for provincial innocence as an Arcadia of individualism, whose power not merely to expose but to *convert* the devious and the corrupt has an almost miraculous quality. The roots of Capra's heroes are perhaps in the Bible Belt tradition of popular American religion rather than in any specific political ideology, and their combination of folksy ingenuousness and charismatic influence can be seen as betraying the hunger for an individual national saviour so important in the rise of rascism in Europe. The films are a fascinating insight into the dangers of 'personality cults' at times of economic stress, as much as into a corruption whose workings they nowhere adequately analyse. If it was possible in the thirties to speculate seriously on the dangers of an American form of Fascism, Capra's work helps to show why.

Capra's Arcadia is a provincial, small-town one; but the concept is at work elsewhere in the American cinema of the thirties. It underlies the ambiguous sub-genre known as the 'screwball comedy', predicated upon rivalry between the sexes, involving the characters in wildly eccentric behaviour and dizzy defiance of social conventions, and almost invariably set in an affluent upper-class milieu. Cary Grant, James Stewart, Katharine Hepburn and Ann Sheridan are among the leading performers in the genre, George Cukor, Howard Hawks, Preston Sturges among its leading directors.

Screwball comedy has recently received a good deal of serious critical attention, notably from feminist critics who see in Hawks's comedies in particular an outlet for sexual insecurities and anxieties which American society could neither articulate more 'seriously' nor disavow altogether. This is certainly a very important aspect of it, especially in the films in which Hepburn and Grant are played off against each other. If Hepburn was for a long time regarded as a curse at the box-office, it could well

American Society in and through its Cinema

have been because of her obvious independence and intelligence – unlike that of say Marlene Dietrich, bereft of any consciously manipulative quality. (Her off-screen character, educated, articulate, and not dependent on one bourgeois marriage or, like many of her contemporaries, a fraught string of them, went to reinforce this perception). Grant's urbanity tacks between insolence and near-hysteria, whether as the withdrawn paleontologist in *Bringing Up Baby*, forced into drag and what may well be the first use of the adjective 'gay' in something like its modern sense, or as the upper-class wastrel in *The Philadelphia Story*, redeeming himself and his ex-wife in a celebrated last-reel *coup de théâtre*. There are moments – such as the drag scene in *Bringing Up Baby* or virtually the whole of her/his long struggle against the bureaucratic codification of sexual identity in *I Was a Male War Bride* – when Grant's ability to draw laughter is quite clearly linked to his ambiguous dramatization of an insecurity connected with the Freudian castration complex.

Where then is the 'Arcadian' quality in these most deceptively serious of screen comedies? It lies partly in the shielding of the characters from the harsh economic realities of post-Depression America. Admittedly, by the time *The Philadelphia Story* was made in 1940, the New Deal had eradicated the worst effects of the Depression; but the characters in that and other screwball films come from a social stratum sufficiently affluent hardly to have been touched by it. Robert Sklar in *Movie-Made America* points to the invariably happy resolution of screwball intrigues as offering proof of 'how funny and lovable and harmless the rich could be'.[11] With 'harmless', perhaps, one can take issue; a social class that (even unwittingly) confuses a tame pet leopard with a savage escaped one (*Bringing Up Baby*), or transports its matrimonial insecurities into the supposedly dignified surroundings of a courtroom (Spencer Tracy and Katharine Hepburn as married lawyers on opposite sides of the same case in *Adam's Rib*), is clearly capable of wreaking a good deal of havoc, and if the films were really 'harmless' their charge of sexual ambiguity would be effectively nullified. But it is equally true that the elaborately zany rituals which take up most of the characters' time would have been possible only for the relatively idle rich. Modern audiences, viewing the films largely for their *mise-en-scène* of female assertiveness and male insecurity, may well tend to forget that for the audience for which the films were made – post-Depression, pre-popularized-Freud – their role as glossily reassuring escapism would have been far more important.

This is not to say that class conflicts, notably between earned and unearned income, do not figure prominently in the screwball genre. Cary Grant in *Bringing Up Baby* is a hard-working professor of paleontology (conforming very much to the professorial stereotype, complete with spectacles and absent-mindedness), Katharine Hepburn a wealthy heiress, with the effortless elegance of one who has never had to work for a living (she sinks an implausibly long putt on the golf course without batting an eyelid). The conflict in *The Philadelphia Story* is opened out by the presence of James Stewart as a muck-raking journalist who would rather be writing fiction, and who on the eve of Hepburn's second wedding vies with her fiancé and her ex-husband (Grant) for her affection. The fiancé represents the new industrial bourgeoisie (he has made his wealth through mining, several times referred to disdainfully as in every sense a low form of life), Grant/C. Dexter Haven the feckless last gasp of unearned (and swiftly squandered) income. The class-lines in this film compare interestingly with those constructed a few years before by Jean Renoir in *La Grande Illusion* and *La Règle du Jeu*; but, where Renoir's 'resolution' leaves all sorts of possibilities open, Cukor's (based on a successful play by Philip Barry) is far more homogeneous. Hepburn's fiancé is never seriously in contention while Stewart (at whom she throws herself in a drunken confusion reminiscent of Christine in *La Règle du Jeu*) is too much of a gentleman either to take sexual advantage of her or to continue writing for his scandal-sheet, and it is left for Grant, with aristocratic aplomb, to take over wedding and bride (for the second time) at the end. This can be translated (approximately) as follows: the *nouveaux riches* lack style (here an ethical, almost a spiritual, judgement as much as an aesthetic one); the lower middle-classes are decent, but fundamentally too decent to be really interesting, to say nothing of the demeaning ways in which they have to earn a living; 'aristocratic' male flair alone can prevail over the independence even of a Katharine Hepburn and get her to do what she has really wanted all along. The film's appeal from 1940 to the present day is, I suspect, largely based upon its 'moral' that adroitness and repartee can beat filthy lucre at its own game.

In *Bringing Up Baby*, the 'aristocratic' scale of values is represented by the woman, who at the end gets what she wants with little apparent sacrifice of independence. On the other hand, the resolution takes the conventional form of marriage (not actually shown as in *The Philadelphia Story*, but clearly inevitable), and the film's anti-intellectual bias is inescapable. What else is Cary Grant's brontosaurus, laboriously reconstructed only to be destroyed by Hepburn's irruption at the end, but a representation of the academic world and its values as, literally, ossified? Hepburn is boisterous as upper-class females in Anglo-American culture have always been entitled to be ('jolly golf clubs' here rather than 'jolly hockey sticks'), but ultimately 'all woman'; Grant's desiccated scholastic intelligence requires the tuition of Hepburn's scatty elegance so that the perfect marriage of 'male' and 'female' qualities can take place. And, without the privileged situation of both characters (Hepburn's unearned wealth, Grant's professorial freedom from normal working constraints), the intrigue would be literally unthinkable.

This is an opportune moment to deal with two possible objections to my analysis. Some may find it portentous, turning major comic triumphs into sociological thesis-fodder rather as Grant's professor turns life in all its variety into a skeleton for analysis and reconstitution. To this I would say, first, that it is not privileging humour, but rather treating it as implicitly second-rate, to decline to look closely at its functioning and the implications behind it; and, secondly, that it is precisely the unquestioned status of these films as comic masterpieces of their time, whose impact is still considerable today, which makes it important to look at them closely in the context of a study such as this. The aspects of itself a society chooses and is able to present as laughable are among the most important in studying that society's representation of itself.

The second objection may be that I have somehow 'criticized' *The Philadelphia Story* and *Bringing Up Baby* for being insufficiently radical in their social comment. This impression is easy to give when analysing a 1930s or 1940s film from a 1980s perspective, but is misleading nevertheless. We have already seen that films carry inscribed within them the laws of their own commercial profitability. This means that at different periods different aspects of the films will tend to be foregrounded by their audiences. The screwball comedies I have analysed contain some acceptance of the status quo, and some overt social criticism (as in the unlikeable character of Hepburn's fiancé in *The Philadelphia Story*), but also some oblique comic highlighting of problems and insecurities impossible for a commercial medium to tackle 'seriously' (Grant's sexual identity-crises), which now appears dated or limited, but which was inevitable at the time the films were made. Hepburn's long liaison with Spencer Tracy was carefully cosmeticized by contemporary publicists and writers as a 'friendship' or 'companionship'. If nowadays this seems coy, if the marital resolution of *Bringing Up Baby* or *The Philadelphia Story* irritates modern 'emancipated' audiences, this historical perspective, with Hays breathing down artists' and studios' necks, should be borne in mind. It could even be argued (though I would not wish to press this too far) that the special 'greatness' of the two films, and their concomitant richness of appeal to a wide range of audiences, is a factor of the way in which they embody both what American society of the thirties and forties could acknowledge about itself and what it could not.

In most of the films we have discussed so far, women either do not need to earn a living (Hepburn in the screwball comedies) or are so presented as to make their means of support irrelevant (has anybody ever wondered whether Fay Wray was paid equally with her male counterparts for taking part in the expedition in *King Kong*?). The exceptions are, of course, the Berkeley/Warner musicals, but these also constitute an exception within the world of the thirties musical, which otherwise tended to be concerned with its 'girls' as objects for the cynosure of male audiences. The two major genres with which we have still to deal, the Western and the gangster movie, treated their women as decoratively functional accessories. How was the economic place of women represented in other areas of Hollywood in the thirties?

Their bodies were commercializable. That almost goes without saying as part of the phenomenon of female stardom, but it was given an ambiguously mocking inflection by the performances of Marlene Dietrich and Jean Harlow. In *Shanghai Express* and *Red Dust* respectively, they are quite clearly playing prostitutes; but the edge is taken off the portrayal by the studio-exotic locales (China, Malaysia), the 'respectability' of their lovers (Clive Brook, Clark Gable) who in the end redeem them by marriage, and the actresses' constant self-distancing from the parts they play. This last point may seem contentious; Dietrich's well-publicized liaison with von Sternberg (which earnt her a $500.000 lawsuit from his wife), and Harlow's *bon mot* 'Every morning when I get up, I feel a new man' hardly squared with the prevailing morality of their time. But the self-mockery and arch repartee that characterized these episodes and the actresses' personas as a whole is a long way removed from the realities of prostitution. Dietrich and Harlow sold their celluloid body-images, but in such a way as to make it difficult for any audience to accept that they could actually sell their bodies.

The examples of Dietrich and Harlow show that the movies in the thirties were one place where women could answer back, and, more generally, where men had no verbal monopoly. One interesting representation of this is the number of female journalists featured, though as Marjorie Rosen points out in *Popcorn Venus* this was somewhat out of touch with the reality;[12] the massive recruitment into the profession in the twenties slumped in common with most other areas in the following decade. But it still throws an interesting light on the place allocated to women in the America of the thirties. This was, of course, subordinate, and in a more than just metaphorical sense, as is shown by the number of American schools which made married *female* teachers ineligible for continued employment in the Depression and post-Depression years. Why should the screen journalist appear to be a special case?

Primarily, of course, because she uses her femininity to tease stories out of males who would remain obdurately silent with members of their own sex. Secondly, because the world of investigative journalism in particular is a microcosm of the capitalist ethos of competition, in which individual skill and guile allied to judicious use of the cheque-book are required for the triumph of the fittest. A woman able to compete in so 'masculine' a world is almost by definition easier for the men within it to accept on their own terms (the example of Margaret Thatcher's position in the British political establishment is a useful parallel here). Thirdly, because in many films there is a sense in which the 'story' the female journalist is seeking is really the resolution of her own 'story' via romantic union with a male; the two 'stories' often coincide so that the apotheosis of her career as a reporter is overshadowed by the beginning of her career as a wife.

These three points all emerge clearly in *Mr Deeds Goes to Town* (Jean Arthur reporting on Gary Cooper's good fortune), and in Howard Hawks's *His Girl Friday* (with Rosalind Russell as part-resentful employee of Cary Grant). Arthur is actually assigned to beguile the innocent Deeds with her femininity via a carefully-staged 'accident', Russell is mandated by Grant to cover a jailbreak because of her enticing skills. Arthur does her job so well that Cooper/Deeds falls in love with her (the ultimate accolade to her professionalism), while Russell exchanges lightning badinage on terms of at least verbal equality with Grant in probably the screen's fastest comedy. Arthur sacrifices professional to personal success and opts for Deeds the man rather than Deeds the story, Russell starts the film as Grant's grudgingly loyal aide on the newspaper and ends it, the story safely lodged, as his cheerfully loyal aide for better, for worse. Whether the news story is sacrificed or merely subordinated to the resolution of the woman's story through romantic union, the result is substantially the same. Her independent career comes to be read retrospectively as a staging-post on the road to love and marriage, and the search for news and information – her life in the public domain – has been but an alibi for the deeper search for 'good news' in the private domain.

One of the few women to win recognition as a major director in Hollywood, Dorothy Arzner, tackles in two of her better-known films the problems of women attempting to make a career in a male-dominated world. In *Christopher Strong* her central female character – an aviatrix played on her second screen appearance by Katharine Hepburn – uses her career as a way out, via suicide, of the dilemma she finds herself in when pregnant by the prominent politician after whom the film is named. She embarks upon a supposedly record-breaking flight, but deliberately flies too high and crashes to her death. Once we have fought our way past the comically stilted studio representations of London and the stiff-upper-lip dialogue, the image of the solo flight as emblem of a woman's independent career reveals itself as quite powerful; but certain uneasy questions remain. Why does the film explicitly equate career success with monogamy in a man and virginity in a woman (Strong and the aviatrix are brought together as the result of a treasure hunt at a party whose goal is to find a man who has never been unfaithful to his wife and a woman who has never had a love-affair)? Why does it suggest that a 'solo flight' towards an independent existence is likely to end in disaster? Why is it named after the central *male* character?

Part at least of the answer, of course, is that Arzner at this stage in her career needed to keep a wary eye on the sensitivities, not just of her audience (the male title may well have been intended to designate the picture as a 'woman's film' and attract female spectators), but of the overwhelmingly male-dominated Hollywood establishment. In *Dance, Girl, Dance,* made several years later, these considerations seem to have been less pressing. The film dramatizes the multiple conflicts, personal and professional, that mar the friendship between two dancers, Lucille Ball/Bubbles and Maureen O'Hara/Judy. Judy, a would-be ballerina, is forced through misadventure into dancing in a night-club; her colleague Bubbles (a rather more sophisticated version of the girls in *Gold Diggers*) has been doing this for a while, partly because it is a reasonably effortless way of making a living but mostly as a means of meeting a wealthy husband. The film's narrative resolution gives both women satisfaction, Bubbles when she finally secures a wealthy husband (whom she also happens to love), Judy when her ballet-dancing talent is rediscovered.

But before this happens there have been two major confrontations: one between Judy and the leering male audience in the night-club where she dances, the other – a result of the first – a public fight between Judy and a Bubbles irate at being upstaged. Judy, whose ballet routine has earlier been used to mock the audience and whet their appetite for Bubbles's vamp number, rounds on them and berates them vigorously for their squalid voyeurism in what has become a classic feminist sequence. But, as is pointed out in *The Work of Dorothy Arzner – Towards a Feminist Cinema*, the wild applause with which this outburst is greeted largely negates its substantive impact. It is reintegrated into the spectacle whose harmony it has disrupted as a particularly *spectacular* example of female tigerishness, just as the fight that follows has an obvious effect of sexual arousal. The likelihood is that *Dance, Girl, Dance,* if only through its title, would have attracted a more male-dominated audience than *Christopher Strong*; and how far that audience would have been titillated, rather than chastised, by Judy's outburst it is very difficult to say.

The competitive ethos once again presides over the happy resolution of both women's stories. Bubbles is clearly far more talented than the average night-club dancer would have been, and the measure of Judy's ability is taken by her 'Pygmalion' ballet-tutor at the end when he says: 'She was born with more than any dancer we've got and she knows less. It's our job to teach her all we know.' The 'best' triumph (the rest, presumably, will continue to jostle or vegetate in mediocrity until they learn their competitive lesson), and their triumph takes the form of acceptance – personal or professional – by the male members of society who hold all the (financial or cultural) capital. In America of the thirties and forties, it could hardly have been otherwise.

Nothing since our discussion of *The Birth of a Nation* has suggested that there might have been inhabitants of the United States who were not white heterosexuals. The portrayal of the Red Indian will be dealt with when we come to discuss the Western; the Negro, no longer representable as a sexual threat or temptation after the furore over *The Birth of a Nation*,* appeared largely in grinning servant roles of the type made famous by Stepin Fetchit. What is most offensive about these is their implication that the Great Society had dissolved all racial and cultural antagonisms in its melting-pot, that the coloured servants were happier serving their white masters than they could ever have been otherwise. The comment of Hattie McDaniel (many of whose roles actually work within the 'mammy' stereotype to subvert its submissiveness by truculence, even rebellion) is at once poignant and unanswerable: 'Why should I complain about making $7000 a week playing a maid? If I didn't, I'd be making $7 a week actually being one!'[13]

As for homosexuality (still illegal in certain states at the time of writing), it did not begin to appear even in a veiled form until the forties (Peter Lorre/Joel Cairo in *The Maltese Falcon* could well be among the first recorded sightings). But homoeroticism – the preference for the company of one's own sex, and the sense of greater emotional ease in male company – pervades the Western (and, to a certain extent, the gangster movies) to a degree

---

*This at least is the theory of Donald Bogle in *Toms, Coons, Mulattoes, Mammies, and Bucks*.

that would have horrified its macho stars had they been aware of it.

The Western and the Gangster Movie

Before we move on to look at the image(s) of America produced by these two key genres, it is worthwhile examining the major studios' tendency to specialize, and hence to divide up the market among themselves, which became so important in the thirties. There was an element of chance in this, since it depended partly upon which stars a particular studio happened to have under contract; but other factors were more important. Critical stress on the director as creative source or *auteur* of a film has minimized the crucial role in this period of the individual producer, whose personal tastes, entrepreneurially harnessed to the bidding of the market, could shape the whole image of a studio and a genre. The most obvious example is Arthur Freed, who became a producer for MGM in 1939 and shaped the reputation of the studio and the musical through into the sixties.

The political *parti pris* of a studio's chief executive could also play an important part (for instance, the already-quoted example of Jack L. Warner's support for Roosevelt and the-studio's social-populist image in the thirties), and so obviously would its financial situation. MGM, incontestably the wealthiest studio, was able to make costly musicals and period dramas in which no expense was spared on accuracy of detail. RKO, on the other hand, bankrupted in the aftermath of the Depression and after the Second World War encumbered with the wayward and financially unsound presence of Orson Welles, were glad of the budget horror movies produced by Val Lewton in the early and middle forties. *Film noir*, likewise, can be seen as a reaction to financial constraints, its sombre black-and-white lighting and cut-price stylization of urban squalor producing maximum atmosphere at minimum outlay.

Steven C. Earley's *An Introduction to American Movies* helpfully summarizes the type of genres in which the different studios specialized[14] — sophisticated 'European' comedies for Paramount, populist social drama for Warner, horror for Universal, cut-price Westerns for Monogram*, lavish musical spectacle for MGM, and so forth. The Western and the gangster movie, however, are very difficult to pigeonhole in this way, for virtually every studio had its specialist director(s) in the two genres, and many of their leading practitioners migrated from studio to studio (John Ford, for example, who worked successively for United Artists, RKO, and Warner Brothers). Part of the reason for their universality has been suggested in Colin McArthur's view of 'America talking to itself about . . . its agrarian past, and . . . its urban technological present'. What did America have to tell itself in the thirties about the agrarian past?

That past, of course, was by no means as distant for an audience then as it is now. The New Deal included important grants and incentives to farmers, the Tennessee Valley Authority provided hydro-electric power to irrigate and open up an area much of which had hitherto been an isolated dustbowl, and we have already seen the depth and resonance of the reaction against cities and towards the countryside that followed the Depression. The 'agrarian past' and the opening-up of new frontiers would have been in the post-New-Deal era realities to be revaluated and rewritten in the light of the movements mentioned above.

But for most modern (especially non-American) audiences, the Western is so steeped in myth as to be effectively ahistorical. When as a schoolboy I consumed a copious diet of television Westerns, it never occurred to me that their action would have had to take place between 1836 (the year of the Alamo and Texan independence from Mexico) and the First World War, if not 1890 (when the Wounded Knee massacre took place and the 'frontier' was officially declared obsolete by the Superintendent of the Census).[15] They were simply 'Westerns', and as such immune from habitual constraints of time and place. This impression may well account for the wide currency of attempts to codify the possible range of Western plots, notably that of Frank Gruber with its seven basic types. The problem with such a taxonomy, apart from the fact that it actually does very little to help read any individual Western, is that it contributes to the mythical atemporality of the world in which a man's always got to do what a man's got to do and sunsets exist for heroes to ride off into. In fact the Western is pervaded just as much as any

---

*It is to this studio that Godard's *A Bout de Souffle* is dedicated, in honour of its low-budget innovativeness.

other genre by the social determinants of the time at which it was made – a point developed (sometimes stimulatingly, sometimes turgidly) by Will Wright in *Six-Guns and Society*.

Wright sees the 'classical' Western (earliest quoted example: *Dodge City* of 1939) as closely linked to the competitive market economy. Its main narrative ingredients are identified (to summarize) as a lone stranger riding into a community, overcoming mistrust to win recognition, helping the community overcome a violent threat from outside, and thereby winning an acceptance which he may either himself accept (by marriage and integration) or have the plot refuse for him (by death or departure). The connection between this and the liberal market economy is clear enough, through the stress on self-reliance, the precarious and not necessarily binding, but voluntarily entered into and maintained, relationship between hero and community, and the notion that individuals and societies deserve to survive only in so far as they can defend themselves or recruit defenders on the 'labour market' of the open plains.* But as often in Hollywood this ideology periodically 'cracks', and the unevennesses and contradictions thus produced are no more than we should expect from works made at a period when the philosophy of competition had had to be bailed out by state intervention. Errol Flynn/Wade Hatton, the hero of *Dodge City*, moves during the narrative of the film from railroad foreman to cattleman to (initially reluctant) sheriff, leaving Dodge City with a church and a choir to compensate for the abolition of its guns, gambling-dens, and saloons. But he leaves at the end for an apparently even more demanding assignment in Virginia City – because he cannot refuse a society that needs his help, or because cities without violence and vice to eradicate are unacceptably boring places for Western heroes (and presumably hell on earth for an Errol Flynn in particular)? The ambiguity casts light on the whole equivocal relationship between self-interest and the good of society, unproblematically fused together by political apologists for market capitalism.

John Ford's *Stagecoach*, made in the same year as *Dodge City*, abounds in paradoxes and anomalies that make it a very rich text for analysis. It is seen by Wright as a 'vengeance' Western (John Wayne/the Ringo Kid is motivated by the desire

---

*This sub-theme is explored in the Japanese *The Seven Samurai*, remade as the Western *The Magnificent Seven* six years later.

to avenge – hence symbolically to annul – his father's death at the hands of the Plumber brothers*) in which the hero's need to avenge causes him to leave, or be forcibly extruded from, a society into which he reintegrates himself at the end through abandoning the drive for individual retribution and dealing with the villains on behalf of society. This clearly reflects a less aggressively individualist approach and a greater tendency towards social co-operation, and as such is linked by Wright to the tentative beginnings of economic planning. While such a view is certainly applicable to *Stagecoach*, the society of the film is largely composed of outsiders. This is clear from the start of the coach's journey across territory menaced by the Indian chief Geronimo; anybody taking part in so dangerous an enterprise is in some sense by definition marginal. The marginality of each of the six travellers is worth a brief analysis.

Gatewood the banker is leaving town with the proceeds of his bank before his fraud is discovered; on the journey he reveals himself in a progressively more odious light, but arrives at his destination in one piece. Doc Boone, like Gatewood, is a 'disgrace' to a supposedly honourable profession (he is an alcoholic doctor); but his 'disgrace' is both more public (he is expelled from town), and less harmful (we are not told that his patients have suffered as a result of his drinking), and he redeems himself professionally by sobering up to deliver Mrs Mallory's baby. Mrs Mallory is the soul of respectability (signified by her being an Easterner), but at the same time doubly exposed by being the only 'lady' on board and pregnant to boot. Peacock is a whisky-salesman, so connected with the saloon world from which Dallas and Doc Boone both stem, but quietly obliging in his manner, and not really too troubled by the number of his samples Doc Boone pours down his throat in the earlier part of the journey. Hatfield, as much the perfect gentleman in appearance as Mrs Mallory is the perfect lady, is in fact a professional gambler; his gentlemanliness and his gambling are reconciled when he protects Mrs Mallory and sacrifices his life during the Indian attack. Finally, Dallas, the 'saloon-girl' (Hayseee for prostitute), bears a double mark of shameful exclusion; like Doc Boone she is forcibly evicted from town, like Hatfield only more so she lives

---

*The felicitous irony of the brothers' name, in the light of Wayne's stubborn post-Watergate defence of Richard Nixon, is strictly irrelevant, but richly worth pointing out.

by a profession foreign to the Puritan ethic, whose embodiment is the Ladies' League responsible for her expulsion.

It is thus perfectly consistent that at the end of the film Dallas and Ringo – its two most unequivocal outsiders – ride off together, and Doc Boone's celebrated *envoi* to them ('They're safe from the blessings of civilization') derives its irony from being uttered by one all too familiar with 'blessings' that neither Ringo nor Dallas have ever, for the film's purposes at least, known. The profession of supposedly legitimate acquisition – banking – and that regarded by the Puritan ethic as second only to theft – gambling – are contrasted to the advantage of the latter. Legitimate and illegitimate use of alcohol come together when Peacock provides Doc Boone with just enough whisky to keep him happy without incapacitating him for unexpected obstetrics, as do legitimate and illegitimate sexuality when Dallas helps to deliver Mrs Mallory's baby. All these various oppositions are in some way connected with that between society and the outlaw, gloriously transcended at the end when Ringo gives himself up after killing the Plumbers, to be set free by the sheriff Curly.

On one level, then, the film operates as a vicarious escape from the constraints of the Puritan ethic, proving that the drunkard, the whore, and the outlaw are not necessarily worse people than the rest of us. Looked at more closely, however, it can be seen as posing less of a challenge to American respectability than this implies. The more apparently undesirable characters are, the more they are shown as having reasons for this in their pasts. Dallas has been forced into prostitution by the death of her parents in a massacre (a convenient strategy for deflecting blame for the downfall of virtue onto the 'savage' Indians). Ringo's obsession with vengeance stems from the cowardly murder of his father; by facing and killing the Plumbers like a man, and doing so in the name of group solidarity rather than individual bloodthirstiness, he not only avenges his father's death, but aligns himself with the forces of law which are then able to pardon him. We are not told why Doc Boone drinks so much nor why Hatfield has elected to make a living by gambling, but, since the doctor is able to deliver the baby and Hatfield to stake his own life with equanimity, it hardly matters; both have redeemed the flaws in their profession or their exercise of it. The only character whose motives remain unclear and whose actions do not redeem him is Gatewood, which may be why I find him the most interesting. *Stagecoach* can be read either as a denunciation of the hypocrisy of small-town America or as a pious plea for understanding and a half-promise that, if only this is shown, even the most apparently unregenerate will turn out to be fundamentally in agreement with the liberal society of goodwill and its values. Or rather, it is both of these at once, which is why it is particularly interesting as an example of how the United States reconstructed its 'agrarian past' for the benefit of its 1939 present.

What is certainly true of the Western as a genre until such comparatively recent examples as *Little Big Man* is that its elements of atemporality or nostalgia work to blunt even the suggestion that political and ideological forces are at work in it (with exceptions, such as the undeveloped hint of financial corruption in Gatewood). By and large, Indians are savages whose pursuit of the white man is ascribed to primeval bloodlust; drawing on personal experience again, I can never recall questioning why the Indians behaved in so rowdy and unreasonable a manner in the television Westerns of my youth. They were Indians, and thus there to behave like that. There were exceptions, such as Alan Crosland's *Massacre* of 1935, which severely criticized the fraudulent and often sadistic treatment of Indians on reservations. But this was not what the public, or those who contributed to form its image of America's past in its present, wanted.

Whatever the merits of a schema as richly suggestive as that of Wright in his *Six-Guns and Society*, it needs to be borne in mind that the correspondences he proposes were teased out by highly sophisticated analysis up to thirty-five years after the films began to circulate. Along with their rejection of the frequent hypocrisies of urban civilization and their compensating stress on the need for self-reliance tempered by humanity and understanding, what *most* audiences from 1939 to the present day would be most likely to absorb from *most* Westerns would be a scale of values in which it would appear immemorially natural that the white man is superior to the Red Indian on every possible level, to the white woman on grounds of toughness and reliability (though he will need to water himself occasionally in the saloon of feminine sympathy), and to the Negro simply by virtue of existing in a landscape where no black faces are to be seen, despite the high number of Negro cowboys who actually took part in the opening-up of the West. The subversiveness of the best examples of the genre (illustrated by our analysis of *Stagecoach*) can still, in its unreflective notions of superiority allied to its populist suspicions of corruption in any and every political organization, come disconcertingly close to the Fascism thought to be a real possibility in the America of the early thirties.

The gangster film, of course, was much more overtly rooted in the society of its time. Gangsterism as a historical phenomenon was the contradictory product of the Puritan and market-capitalist ethics. The former led to the prohibition of alcoholic beverages throughout the United States in 1919, a move that would surely have gladdened the hearts of Ford's Ladies' Legion, but did nothing to stop the development of such spectacular Hollywood drinking careers as those of F. Scott Fitzgerald or W. C. Fields; the latter ensured that the consumers' wants continued to be satisfied, and extended the market precepts of free individual action and unbridled competition into the realm of homicidal violence (the notion of the state as umpire seems to disappear from the model at this point). The best market justification for Prohibition is probably that it provided an alternative (and financially rewarding) career-structure for many of the poorer sections of society; Al Capone, like his cinematic alter egos Rico in *Little Caesar* and Tony Camonte in *Scarface*, came from a poor Italian immigrant background.

The first gangster films nowadays regarded as important – the two just mentioned plus *The Public Enemy* – all date from between 1930 and 1932, a transitional period in the history of racketeering. Prohibition was to come to an end just a year later, and industrial protection and extortion, along with the occasional kidnapping (such as the Lindbergh affair), were to replace it as the main sources of gangster income. American trade unions have always been a byword for corruption and brutality; indeed, if such fifties and sixties gangster films as Fuller's *Underworld USA* were to compare the underworld to a highly efficient big-business concern, Elia Kazan in *On the Waterfront* was in a sense to do the reverse, by constructing a dockland union organization rife with corruption and resorting to gangland tactics as a matter of course.

This shift of focus in both real and cinematic gangsterism (not that the two necessarily went hand in hand) was accompanied by a shift in the depiction of lawmen, who until the mid-thirties had generally been corrupt, obtuse, or both. There were several reasons for this: the arming of FBI agents meant that they could now be shown pursuing villains much as cowboy heroes pursued rustlers, Indians, and other undesirables; civic pressure-groups, with the backing of the Hays Code, accused the major studios (not without reason) of glamorizing robbery and violence, and of therby having a pernicious effect on the young in particular; and the shift of focus from Prohibition to labour racketeering removed the overt workings of gangsterism from the public eye.

As examples of the change in gangster films that took place in the mid-thirties, it is worthwhile analysing Wellman's *The Public Enemy* of 1931 and Curtiz's *Angels with Dirty Faces*, made seven years later. Both films star James Cagney as the gangster figure (Tom Powers in *The Public Enemy*, Rocky in *Angels with Dirty Faces*), and in both he is counterposed to a 'good' character (his ex-serviceman brother and Pat O'Brien's priest respectively). *The Public Enemy* goes some way towards recognizing that poverty and inadequate housing can contribute to crime, but the fact that Tom Powers becomes a Prohibition racketeer while his brother remains on the right side of the law is not explained or analysed. Furthermore, the Cagney character is much the more vital and energetic of the two (his brother is tediously sententious), to say nothing of his frequently displayed love for his mother ... There is little doubt where the focus of audience identification would have been, and the narration of Tom's criminal career also goes to undercut the rather crude portrayal of socially determining factors. For it is constructed as just that – a *career*, in which we watch his progress from apprenticeship (the bungled adolescent burglary at the beginning) to the top of his 'profession'. His death at the end – deposited swathed in bandages on the doorstep of the mother he adored – is only an apparent contradiction of this; because, as McArthur notes, 'that the gangster must eventually lie dead in the street became perhaps the most rigid convention of the genre',[16] the ultimate accolade for the successful gangster became eradication by rivals for whom he has come to represent too much of a danger. (*Little Caesar*, made the year before, has a 'rise-and-fall' structure which means that Edward G. Robinson/Rico finishes the film as a drunken derelict; between his shooting and that of Tom

Powers, apparently similar, there is a world of difference in 'career' terms.)

The uprightness of Tom's brother Mike turns the 'social-problem' aspect of the film back upon itself. Why should one brother from a deprived background turn to violent crime while the other rejects it, even to the extent of disrupting a family celebration by refusing to drink Tom's bootleg beer? Unless, of course, 'the child is father to the man', which makes the poor background a pretext for an implicit sermon on the innateness of good and evil.

This reading is reinforced by the film's insistence on Tom's sexual abnormality – not just his exaggerated devotion to his mother (index and product of his Irish-Catholic background), but his sadism towards women, exemplified in the notorious scene where he smashes a grapefruit into Mae Clarke's face at the breakfast-table, for no better reason than that she has just tearfully asked him if there is somebody else in his life. Rico in *Little Caesar* is, at least latently, homosexual; Tom's near-psychopathic violence can, like homosexuality, be read as the product of an exaggeratedly matriarchal culture; but this, like the deprivation theme, is nowhere explored or developed, merely left lying around in the film for audiences to decipher as best they can. More sophisticated audiences at the time of the film's release, like modern ones familiar with Freudian concepts, would have had little trouble in reading Tom's violence and mother-worship as two sides of the same coin. Popular audiences of the thirties, however, would have been most unlikely to have done so. To them, Tom's conduct would have appeared bewilderingly schizophrenic, the index of an almost transcendental malignity. Like the social background, the theme of psychopathology can work to reinforce the sense of innate evil rather than to explain or neutralize it.

There is far less scope for audience identification with the gangster-figure in *Angels with Dirty Faces*. As in the earlier film, two youths who have grown up together in the same deprived neighbourhood follow divergent paths – more divergent in one sense (Pat O'Brien is a priest, whereas Tom's brother in *The Public Enemy* is an ex-serviceman with an ambivalent attitude towards killing), in another sense less so, for when Rocky returns to the scenes of his childhood and becomes the idol of the local youth his relationship with the priest remains a friendly one. Pat O'Brien is an underrated, but extremely important, figure in Warner's thirties films – to quote David Thomson, 'their resident apologist for the social order, either as cop or priest',[17] but sufficiently warm and lacking in sanctimoniousness to make that social order a credible alternative. In *Angels with Dirty Faces* he alone of the neighbourhood worthies recognizes positive potential in Rocky (notably in the scene where Rocky referees a basketball game and through verbal and physical agility actually succeeds in imposing order on chaos – an ironic reversal of the gangster's usual function). The film's ending is celebrated for its ambiguity; Rocky, captured and sentenced to die in the electric chair, is begged by his childhood friend (and sometime partner in crime) not to go to his death defiant, which will add to his heroic stature in the eyes of the Dead End Kids and thus increase the likelihood of their following him into serious crime. Rocky at first contemptuously refuses, but as we see his silhouette on screen in the film's last moments he begins to struggle in panic and cry out: 'I don't wanna fry!' A redeeming last-minute change of heart, justified by O'Brien's knowing expression and the scornful reaction of the Dead End Kids? Or, in a more deadly twist of irony, a moment of unfeigned terror, placing cowardice rather than deeply buried decency at the heart of the gangster psyche? The ambiguity in any event leaves no doubt that gangsterdom is a fool's paradise, and the ending, whichever interpretation we choose, is a clear symbolic atonement to the forces of decency for the supposed corrupting effect of earlier movies in which gangsters were glamorized. Pat O'Brien's look of relieved absolution at the end can be seen as Hollywood's *mise-en-scène* of its hoped-for forgiveness by the Catholic pressure-groups largely responsible for the industry's auto-censorship. The distance travelled in seven years is considerable, and revealing.

Gone with the Wind

The most legendary Hollywood picture of the thirties, if not of all time, *Gone with the Wind*, has a double claim to analysis in a study of this sort. Its history has in two senses become legend. The Civil War conflicts it reconstructs and dramatizes (in its first half at any rate) acquire any historical (as opposed to mythical) significance they may have only through the discourse of Clark Gable/Rhett Butler; for the rest, the war is a series of epic set-pieces in which the political element is almost entirely evacuated by the cultural, and history is collapsed into the usurping of one life-style by another markedly less likeable. At the same time, the history *of* (as opposed to in) the film has become legendary as the most spectacular example of how Hollywood could project and dramatize its own workings on the grand scale. Three directors (Sam Wood, George Cukor, and the finally-credited Victor Fleming); well-nigh countless scriptwriters; assiduously fostered rumours and counter-rumours about the casting of Scarlett O'Hara, before David Selznick's choice fell upon the little-known Vivien Leigh – nowhere else has Hollywood's ability to construct a massive and gaudy apparatus of highly publicized myth around the competition of individuals for roles and status so clearly demonstrated itself. *Gone with the Wind* would belong in a history of American advertising and publicity as a market phenomenon even if all copies of the film were lost (which of course, is not likely; the master is preserved in a solid-gold box in MGM's vaults).

It is partly, but not entirely, because of the barrage of publicity surrounding the film's stars (would Selznick be able to afford Gable? who would eventually be cast as Scarlett?) that *Gone with the Wind* has gone down in history as a great Hollywood love story rather than as an ambitious reconstruction of the Civil War. The film focuses increasingly as it moves through its second part on the pitched battle of wills between Rhett Butler and Scarlett O'Hara, at the end of which Butler, not giving his celebrated damn, walks out with his freedom, leaving Scarlett alone with the estate of Tara as consolation and – maybe – bait to lure her husband back. Unreconstructed emotional capitalists both, it is hardly surprising that Rhett and Scarlett remain probably Hollywood's most notorious couple; but more interesting from our point of view is the Civil War conflict that dominates the first half of the film. The reconstruction of this owes nothing to realism (the husband of Margaret Mitchell, on whose novel the film was based, said after the première, about the famous crane-shot of the Atlanta dead and wounded: 'If we'd had that many dead, we would have won the war'), and precious little to any sense of the historical issues involved.

Charles Hamilton describes the society of the Southern gentry at Twelve Oaks as 'a whole world that wants only to be graceful and beautiful', and it is style in all its doomed and precarious elegance that defines the South throughout the first half of the film. Closely linked with this, until undercut by Scarlett's machinations, is the solidarity within and between the two families, at Twelve Oaks and Tara, exemplified when Scarlett uses her family's most precious curtains to make a dress with which she tries to seduce Rhett into contributing to the land-taxes.

The intensity of Scarlett's attachment to the land is ascribed by her father to her Irish ancestry – this being only one among the multiplicity of class and/or ethnic stereotypes with which the film's canvas is peopled. The Negro servants are as predictably loyal to their masters as the Yankee carpetbagger Jonas Wilkerson is meretricious, the local whore turns out to have a heart of gold when she offers a substantial contribution to the war-effort, and the indignant local ladyfolk reject this as 'dirty money'. The beginning of the film's second half inverts a stereotype to powerful effect, when the disinherited aristocrats are shown picking cotton in the fields like the 'poor whites' they have so despised; but this, following as it does close upon Scarlett's vow never to be happy until Tara is once more back on its feet, is promptly recuperated into a lesson in the virtues of hard work and self-reliance. The charm and grace of the old order, it is suggested, can be won back only – if at all – by the self-sacrificing toil that must have been necessary to build it in the first place.

Against this displeasing combination of sententiousness, stereotyping, and sentimentality, the only voice to be raised, as mentioned earlier, is that of Rhett Butler. Gable/Butler's bluff, insolent virility acquires a political dimension when set against the effeteness of the local 'beaux' (notably that of Leslie Howard/Ashley Wilkes, whose capacity to rival with him for Scarlett's affections is one of the film's major mysteries). He alone appreciates that the old South is a doomed anachronism and that the concentration of industrial potential and raw materials in the North makes its victory inevitable; his decision to fight springs from perversity (and perhaps sexual exhibitionism) rather than patriotism; and the ruthless individualism with which he applies himself to the task of breaking Scarlett's will is of a piece with his hard-headed capitalist view that the feudal South is doomed. Scarlett's 'Tomorrow *is* another day!' at the end of the film, as 'Tara's Theme' blazes from the soundtrack, has nourished in generations of audiences the hope that she and Rhett will, in some extra-filmic Utopia, be reconciled; but it is Rhett who, in turning his back on the landed heritage and what it represents, has recognized the ironic truth of the words. The Gable style of urbane machismo has dated considerably, but capitalism has had few more persuasive screen advocates than Rhett Butler.

The War Years and *Film Noir*

There is not a great deal in the late thirties films we have analysed to suggest that within a few years of their making the United States would be plunged into a global war. Pearl Harbor had a traumatic effect upon the film industry, for it was not too far away from Hollywood; but most of the films of the early forties that are at all widely available now seem to have little or nothing to do with the war as such. Why should this be?

One major reason is certainly that a great deal of the energy generated by the Hollywood war-effort went into propaganda and instructional films whose life was necessarily short. The short-term success of these – the only valid criterion by which to judge them – can be assessed by the block banning, in December 1942, of *all* American films in France and other occupied countries, and by Walter Wanger's proud description of the film-prints exported from the United States during the war as '120,000 American ambassadors'. Paul Hunter's observations in the magazine *Liberty* on 19 December 1942 are worthy of quotation, not so much because of their risibility as because they demonstrate how central the film industry was thought to be for the communication of a positive image of American society:

> There is no substitute for motion pictures. Neither radio, magazines, newspapers, nor the inspired oratory of our most silver-tongued statesmen can take the place of the living, moving story which unfolds before us in the darkness of the theater (*sic*), enters into our souls, and makes us live and feel

the defiance, the courage, the sorrow, the sacrifice, and the sublime glory which is America at war.[18]

Clearly, the cinema's claims to cultural and social respectability (vitiated in many American eyes by the drink/drugs/sex scandals that had blighted it towards the end of the silent era) were fully reasserted, along with its privileged status as the only art-form that somehow enabled its consumers to 'live and feel' the tale it was telling. Such a notion is still widespread, as the recommendations of the 1979 Williams Committee on Obscenity and Film Censorship, in its report to the British government, show. This described film as 'a *uniquely* powerful instrument' (my italics), on which basis it felt able to recommend a qualitatively different approach towards the cinema from that it wished to see adopted towards other art-forms; censorship of the printed word should be abolished, censorship of the moving picture curtailed or codified, but not done away with.

The common denominator, across the thirty-seven years that link a piece of American journalistic jingoism and the report of a British governmental committee, is a belief that what is shown on the cinema-screen (and which must therefore in some way 'really' have 'happened', whether organized for the camera's benefit like most feature films or ostensibly 'captured' by it as happens with documentary and newsreel footage) has a force – for good or for evil, but at all events for credibility – greater than that available to the photographic image or the written word alone. Such an attitude, if taken as somehow evidencing the superior 'essence' of cinema, is as meaningless as any other statement about the 'essence' of any art-form, and its persistence through into the era of colour television, cassette-recorders, and other widely available forms of cultural diffusion, is at best only an anachronism. But in the 1940s, when the hunger for documentary footage on both sides of the conflict was so great that studio reconstructions and recycled material were sometimes passed off under that label, its strength was understandable. Not the least important effect of the Second World War from our point of view was its confirmation that the cinema was at that time at least a 'uniquely powerful instrument' for producing and sustaining a national image both at home and abroad.

The movies made by Hollywood in explicit pursuance of the

war-effort have rarely survived to the present day in commercially available form, and those that have have tended to do so as minor curios in the output of an established *auteur* (such as Huston's *Across the Pacific* or Hawks's *Air Force*). It is difficult to disentangle the effect of the war upon films of the early forties from that of other social and economic forces, such as the continuing pressures of censorship, the aftermath of the Depression, and the intense competition between studios which led to the programming of 'double-feature' bills and hence the emergence of the B-film as a distinct entity.

All the above elements coalesce in *film noir*, as difficult to define as any other genre but virtually unique in that everybody agrees on what was its first example. *The Maltese Falcon*, directed by John Huston in 1941, inaugurated a string of films whose basic characteristics are fairly easy to recognize: low-budget black-and-white shooting (at least partly attributable to the economic constraints of wartime and the B-movie), the use of urban, preferably night-time locations (which were ideally suited to the type of film-stock and shooting), and a good deal of sexual tension generally expressed in a covert or convoluted form (Hays riding again). Borde and Chaumeton, in their important *Panorama du Film Noir Américain*, see *film noir* as a synthesis of three pre-existent genres: the gangster movie (in which Warner Brothers had specialized), the horror film (Universal), and the detective movie (Fox and MGM). Economically it can thus be read as an (unconscious?) instance of cooperation-via-competition among the major studios, all of whom had several important examples of the genre to their credit but none of whom sought to monopolize it (it was not at the time financially significant enough for that).

The three contributory genres identified by Borde and Chaumeton have in common a sense of tension and menace which is what audiences today associate with the expression *film noir*, as much on the level of plot (often so complex that it threatens to dissolve into unintelligibility) as on that of décor, lighting, and acting style. Borde and Chaumeton associate the rise of the genre in the forties with the violence of the war years, though the aftermath of gangsterdom and the post-Hitler arrival in Hollywood of European directors such as Fritz Lang or Billy Wilder, accustomed to the often bleak and fatalistic stylization of Expressionism, were also important factors.

It was hardly to be expected that Hollywood at this critical time would produce a string of extravagantly jovial divertissements. What is noteworthy about the major *films noirs*, however, is that their gloom is neither puritanical nor unalleviated, despite the economic, political, and censorial constraints under which the industry laboured. Psychoanalysts of almost every school would agree that sexual repression is a mainspring of most 'advanced' societies, and also that repression is not the same as suppression; that which is repressed will return in one form or another, ranging from the dreams of 'escapist' cinema* to the nightmares of Fascism. Wartime inevitably led to a slackening of the old sexual restraints in society: they could not be likewise slackened in society's celluloid representations, but their repression itself became the vehicle of an eroticism all the more potent because displaced, if not downright euphemistic. The lush undergrowth of *film noir* sexuality contains examples of degenerate nymphomania (Martha Vickers/Carmen Sternwood in *The Big Sleep*), homosexuality (Peter Lorre/Joel Cairo and Sydney Greenstreet/Gutman in *The Maltese Falcon*), homicide as focus/stimulus for eroticism (Fred MacMurray and Barbara Stanwyck in *Double Indemnity*), implied male impotence (Glenn Ford in *Gilda*), and one glorious combination of exhibitionism and fetishism, when – again in *Gilda* – Rita Hayworth in the title-role peels off her . . . *gloves* in despairing response to Ford's indifference while singing 'Put the Blame on Mame, Boys'.

Fetishism is a leitmotif in *film noir*, and I do not mean just the overtly sexual kind. The fabulous jewelled statuette which provides title and motivation in *The Maltese Falcon* is (or may be) a fake, despite which Cairo and Gutman are of one mind in their pursuit of it; this deviant kind of greed – deviant because founded on fetishism rather than on straightforward bourgeois acquisitiveness – is echoed and paralleled by Humphrey Bogart/Sam Spade's fascination with Mary Astor/Brigid O'Shaughnessy, in spite – read *because* – of her duplicity. The lift gates closing prison-like upon her, as Bogart ruminatively

---

*I place the term in inverted commas the better to emphasize that it really means the opposite of what it appears to; the whole point of most 'escapist' films is not that they are radically unlike the lives of most of their audience, but that they evoke and articulate aspects of those lives that are otherwise repressed and unavowable.

fingers the dubious statuette, do not represent merely the detective-story resolution predicated upon finding out the criminal; they are as much of an erotic triumph for Bogart as the capture of the falcon would have been for his adversaries.

In *Double Indemnity*, it is the murderous greed of Barbara Stanwyck and Fred MacMurray which provides the erotic impetus of the film. Their rendezvous in a supermarket to discuss murdering Stanwyck's husband for his life insurance is watched over by posters bearing such slogans as 'We Deliver More for Less' – emblematic of their paltry greed, as has often been pointed out, but also highly erotic. The triangle of money, death, and sexuality receives in this film an articulation all the more powerful because its apex, sexuality, had perforce to be given indirect expression.

Charles Higham and Joel Greenberg describe *Double Indemnity* as being 'without a single trace of pity or love',[19] a superficially accurate but doubly deceptive account. Love understood as trusting affection is certainly absent from the Stanwyck/MacMurray relationship, but the corollary of this is their excited fascination with each other's potential for treachery, similar to Bogart's for Astor or Glenn Ford's for Rita Hayworth in *Gilda*. Women in *film noir* are nearly always beguilingly deceitful (it would be interesting to explore how far this is connected with their greater financial independence and the sexual opening-out of wartime), which fuels the male characters' excitement but diverts their need for trusty companionship onto members of their own sex, or alternatively (as in Bogart's relationship with his secretary in *The Maltese Falcon*) onto women with whom there is no apparent sexual bond. The relationship between MacMurray and his boss (Edward G. Robinson) in *Double Indemnity* is an excellent example of this. It is to Robinson that MacMurray makes his final confession (via the office dictating-machine, symbol of their professional bond); it is of him that he speaks most fondly in the soundtrack flashback commentary; and it is between the two men that the film's final, and arguably first, act of warmth takes place, when Robinson (who has constantly had to borrow a light from MacMurray throughout the film) lights his dying companion's cigarette at the end.

'Without a single trace of pity or love', then, is hardly appropriate, and the final cigarette is interesting also as a talisman or fetish, embodying the precarious, and now almost-but-not-quite-shattered, esteem and trust between the two men, by way of a gesture which, in the nicotine-sodden Hollywood of *film noir*, is the lowest common social denominator, and gains much of its poignancy thereby. Women are fine for mad extravagances of doomed passion, but only a man can really light another man's cigarette: that is what the film's final gesture would have suggested even more powerfully to a commercial audience of 1944 than to its modern-day spectators.

If (as Molly Haskell among others has suggested) the neurotic fear of women in *film noir* was rooted in fear of their increasing financial independence, it finds expression in a very distorted form (not surprisingly considering that distortion is the hallmark of the genre). The women in the films mentioned hitherto are either involved with crime (Mary Astor/Brigid O'Shaughnessy), or financially dependent on husbands they neither love nor trust (Rita Hayworth/Gilda, Barbara Stanwyck). The outstanding example of a financially independent woman in the cinema of the forties was Joan Crawford, who during the decade graduated (if that is the right word) from unrequited lover to ambitious career woman, her real-life broken marriages and inability to bear children interacting with her screen persona. Not until the age of thirty-nine did she play a mother for the first time, an index of how thoroughly she was identified with an independence she was often presented as wishing (in vain) only to lose. This is the double-bind in Hollywood representation of the career woman: her career is either an alternative to the home and family she would secretly rather have, or a dangerous distraction from those she actually has. Crawford's first film as a mother, *Mildred Pierce*, is a textbook illustration of this.

Mildred's first marriage (complete with two children, both daughters) breaks up as a result of her husband's affair with another woman, and Mildred resorts to waitressing to keep her daughters in the style to which they are accustomed. This inspires her with the idea of opening a chain of restaurants of her own; cooking in this film is a clear metaphor for Mildred's femininity, for in her first marriage she is constantly in the kitchen and as an independent woman she has an array of others, preparing, chopping, cooking, and serving, at her command. Hollywood liked people to think of its geographical surroundings as a sun-soaked paradise, but its cinematic representations frequently give the lie to this. The Los Angeles of *Double Indemnity* is 'an urban hell of dark streets, desolate railroad tracks, gloomy apartments and office buildings';[20] that of *Mildred Pierce*, sunnier and less claustrophobic (the sea plays an important part in the décor), is barely more pleasant, peopled as it is by financial double-crossers and sexual capitalists such as the two real-estate agents, Jack Carson/Wally Fay and Zachary Scott/Monte Beragon.

Caught between the 'feminine' domain of the kitchen and the 'masculine' domain of property-dealing, Mildred is obviously in a difficult position; the well-meaning coarseness of Wally Fay's advances is an index of the American male psyche's difficulty in dealing with female business and sexual independence. But the film provides a way out of the dilemma; Mildred is 'really' so involved with the male world of business because of her obsessive love for her daughter, Veda, played as an unnerving amalgam of bitch and brat by Ann Blyth. And Veda 'really' only rejects her mother's affection because it takes such an unremittingly materialistic form throughout (ballet- and music-lessons, new clothes, a costly education, the full panoply of Hollywood's use of money to buy love). If Veda progresses with puberty from the merely unpleasant to the positively evil, this is 'really' only because her mother's substitution of worldly goods for affection has corrupted and embittered her (it also offers interesting food for thought on the difficulty Hollywood had in handling female sexuality outside marriage). Thus, taking this reading to its extreme but logical conclusion, Veda's seduction of her mother's second husband (Monte Beragon) and murder of him in a jealous rage are 'really' only a further index of how desperate her search for genuine affection has become, and the film comes perilously close to saying that it serves the doubly broken Mildred (emotionally shattered and financially bled dry by Beragon's manoeuvrings) right for having had the temerity to step out of her first marriage and commercialize a skill which God had intended for her husband's delight alone.*

But she is redeemed, by the double assumption of her rightful status as mother and wife at the end. So tenacious is her passion for Veda that she attempts to take the responsibility for her daughter's crime – an attempt thwarted by the detective in

---

*I refer, of course, to her culinary abilities; but the ambiguity is intentional.

charge of the case. Veda is led off to an actual prison, where she can hardly wreak more havoc than in the metaphorical one in which she has spent most of her life; and her first husband, loyal, penitent Bert (the contrast of whose name to Monte's speaks ethnic volumes), is at hand to escort Mildred from the gloomy shadows of the police station back, we presume, to the role she would have done well never to have abdicated in the first place.

This probably sounds like a muted denunciation of *Mildred Pierce* for implying that a woman's place is in the home, or else; but there is another side to the picture. It was among the few important Hollywood films to focus on the real problems a woman attempting to build an independent career was likely to face. Mildred can never be sure that the male agents and rivals with whom she comes into contact are not out to exploit her doubly, as vulnerable businesswoman and as vulnerable woman *tout court*. The draining effect of a career constructed against such odds on her emotional, as well as physical, energy is skilfully suggested, as is the addictive spiral of hard work to make money and frenzied generosity in the spending of it. One classic male device for denigrating women with independent careers is to suggest that the career is a surrogate for home and family relationships such women are emotionally incapable of having (the second half of the expression 'career woman' is often articulated as though in direct contradiction to the first). Nowhere does Curtiz's film suggest this; Mildred may prove unable to sustain the dual role of mother and wife/businesswoman, but this is at least partly because the males around her are unable to accept her in it, and she is never constructed as incapable of either good management or genuine affection. Her limitations, like the film's, are those of the society of the time and its willingness to concede career independence to women only at the price of emotional stultification.

Men as well as women were shown by Hollywood as compensating in their careers or public lives for failure or inadequacy in the private domain. *Casablanca* and *Citizen Kane*, perhaps the two best-known Hollywood films of the early forties, both make great play with this theme, so common in the cinema of the time as to suggest a vast lack of national self-confidence. In the wake of the Depression and Pearl Harbor, this was hardly surprising, and it also represents a successful attempt by Hollywood to trade on the insecurities and neuroses with which its name has always been associated. Its social and economic Darwinism produced a whole industry dedicated to relaying and amplifying the nervous breakdowns, traumatic affairs, broken marriages and drink problems that were its inevitable by-product – initially via the rival queens of screen gossip, Louella Parsons and Hedda Hopper, and in the fifties by such scabrously inventive magazines as *Confidential*. (Kenneth Anger's *Hollywood Babylon*, unfortunately now available only in a heavily bowdlerized version, was a later camp parody of these.) The industry throve like flies on a midden upon such gossip, ethically distasteful no doubt but eminently consistent with the rules of commercialization. The survival of the fittest in Hollywood came to mean, not just the need for ruthlessness and unscrupulousness to survive, but a compensatory mechanism of success through the scandal-sheet and the gossip-column. The unhappy marriages of Bette Davis and Joan Crawford, the drunken debauches of Errol Flynn, Judy Garland's recurring problems with barbiturates and breakdown, paradoxically made these stars' unhappiness and emotional troubles their major assets. Success at the box-office – most strikingly with Judy Garland – was actually symbiotic with upheavals in private life, and the logic of the market recruited personal anguish into the service of the gossip sub-industry.

This accounts for the importance of the theme of insecurity from the industry's point of view. From that of the American public, it performed a dual function, at once consoling and minatory. Mr/Mrs/Miss Everyamerican could find relief from the traumas and tumult of their emotional life in the knowledge that their favourite stars suffered in much the same way (one touch of nature making the whole world kin), while at the same time the more spectacular scale and greater amplification of Hollywood anguish acted as a warning that money could not buy happiness. It could, however, buy fame, a highly attractive consolation prize; and the ambivalence thus produced towards Hollywood, city of heartache and broken dreams but also of glamour and extravagance, was of vital importance in sustaining its mythical fascination through the decade.

The geographical Casablanca is a long way from Hollywood; that constructed by Michael Curtiz in 1943 is much closer. As mentioned earlier, the theme of neutrality, likely to be perceived only in passing by a modern audience, would have been very important to audiences at the time of the film's release. Humphrey Bogart/Rick's radical credentials (including gun-running for the Spanish Republicans) are impeccable, and if he now 'sticks his neck out for nobody', that is the result not of ideological disillusionment but of personal suffering. The two in fact interact and criss-cross at several points throughout the film, whose undoubted emotional power has led to its specifically political allusions being all too often neglected or underrated. What brings the *grand amour* of Bogart and Ingrid Bergman in Paris to a premature close is the imminent arrival of the German army, and what causes Bogart, in the famous airport scene, to forgo his own emotional satisfaction and persuade Bergman to join her husband (Paul Henreid) on the plane to Lisbon is the realization that her place is at Henreid's side, helping his fight for the Resistance. The sexism of this conclusion (the idea that behind every successful man there is a woman, and that Bergman needs Bogart to tell her what to do) is that of its time. Its final political message is emphasized when the neutrally self-seeking Prefect of Police (Claude Rains) walks back to the airport buildings with Bogart and en route tosses an empty bottle of mineral water into the waste-paper basket; clearly visible on its label is the word 'Vichy' (seat of the Axis puppet government in occupied France). If professional success is often a means of veiling personal pain, commitment to a larger cause can be a means of alleviating it. *Casablanca* has elements of the existentialist political thriller as well as of the romantic melodrama, and the café in which most of its action takes place, a cosmopolitan limbo, is a 'melting-pot' just as much as any specifically American location could have been.

The intersection of screen and 'real-life' personas, such profitable terrain for the gossip industry, was an important part of the Bogart phenomenon in the forties. *Casablanca* brought out the latent tenderness and sentimentality from beneath the cynical, hard-bitten exterior of *The Maltese Falcon*; the next major stage in the development of the Bogart image came with *To Have and Have Not*, during the making of which he and co-star Lauren Bacall began one of Hollywood's most famous off-screen romances. His previous marriage – to the actress Mayo Methot, nicknamed by him 'Sluggy' – had been characterized by drunken public brawls and shouting-matches; that to Bacall (at least taking the publicity machine at face value) was as

American Society in and through its Cinema

idyllic as its predecessor had been traumatic, and the steady mellowing of his screen persona was seen as indicative of this.

Bogart also provides an interesting touchstone of the Hollywood attitude towards drink, an important weapon in its armoury both on- and off-screen in these years. We have already seen how important Prohibition was to the genesis of the gangster movie, and it had played a major part too in the Fatty Arbuckle scandal of 1921. Arbuckle, a leading comedian of the silent era, hosted a party at which bootleg liquor flowed plentifully, and the death of a 'model' (Virginia Rappe), allegedly in Arbuckle's embraces, led to his trial on a charge of manslaughter. Three trials eventually produced an acquittal, but Hollywood was less forgiving; his career was wrecked, and the upsurge of puritanism that led to Hays can largely be traced back to this *cause célèbre*.

Prohibition did not, we have seen, stop Hollywood drinking on the grand scale. John Gilbert, John Barrymore, and Buster Keaton all slid into alcoholism, at once cause and effect of their declining careers. There was a certain tragic theatricality about Barrymore's drinking in particular (he parodied his histrionic persona to hilarious effect in *Twentieth Century*), but in the forties the Hollywood attitude towards drunkenness (off-screen at any rate) underwent a change. The decade produced the screen's best-known treatment of full-scale alcoholism in *The Lost Weekend*, which goes too far only in its delirium tremens set-piece;* but it also produced a quota of hard-drinking stars whose excesses were regarded as entertainingly subversive rather than tragic or pathological. Bogart drank copiously both on- and off-screen, and it is likely that his ferocious consumption of whisky and untipped cigarettes led to his death at only fifty-seven.† But his excesses were taken as proof of what a 'good Joe' he was, as well as being an obvious occupational hazard of the private-eye business (an angle parodied by Elliott Gould's impossible chain-smoking as Philip Marlowe in Robert Altman's 1972 adaptation of *The Long Goodbye*). Similarly, Errol Flynn's craving for alcohol – so pathetically extreme towards the end

---

*The effect of this film on subsequent screen portrayals of alcoholism is interestingly discussed by Jim Cook and Mike Lewington in *Images of Alcoholism*.
†His widow was indeed to object to his silhouette being used in a cigarette advertisement for this very reason.

that he had friends visiting him in hospital bring oranges injected with vodka – was seen as part of his lascivious, roistering lifestyle, presented so that his fans would desire yet fear to imitate it. W. C. Fields, for sheer dogged intensity of consumption probably the king of them all, built his alcoholic – but never drunken – persona into his films and the machinery of publicity about him, so that it became intentionally impossible to disentangle myth from reality. The different styles of excess – Bogart's cynical bar-room companionship, Flynn's hell-raising, Fields's cantankerous misanthropy – were used to preclude or divert an awareness of the problems causing and resulting from excessive drinking, which like emotional upheaval was built into the picture of stardom. Nothing was not commercializable in the logic of the Hollywood marketplace.

Bogart and Fields, different in a host of ways, shared a profound sense of aloneness that characterized many important male performances of the forties. In the world of *film noir*, this was an obvious response to the treacherous complexity of women and intrigues; in that of Fields's ferocious denunciations of American family togetherness, it was a vicarious outlet for male spectators who could fantasize their wives into termagants and themselves into self-defensive cynics. The 'independent' woman of the *film noir* and the wife and mother of screen comedy remained alike incomprehensibly *other* to the acerbic solitude of the lone male.

But it would be an oversimplification to ascribe the sense of aloneness solely to male fear of women. It dovetails in a variety of ways with capitalist wartime America's view of itself – beset by the insecurities we have mentioned, but also curiously fascinated by them. Alcoholic bouts and emotional traumas maintained exactly the right ambiguous distance between stars and a public that yearned but did not dare to be like them. On top of this came the screen revelation that Humphrey Bogart's Rick, or Fields's 'Uncle Bill' in *Never Give a Sucker an Even Break*, were lonely, soft-hearted individualists beneath the rebarbative exterior. Sexual paranoia and economic insecurity apart, such an image was very useful in suggesting to the American public that their capitalists – small-scale cynics like Rick or megatycoons like Orson Welles's Charles Foster Kane – were decent, misunderstood figures beneath the tinsel and the tantrums, and this was done by a thorough manipulation of *knowledge* that extended from the texts of the films themselves to the gossip- and scandal-industries.

The gossip- and scandal-industries acted to ensure that the private griefs and problems beneath the public exterior would not long remain private. Greta Garbo remains unique among major Hollywood stars in having succeeded in retaining her privacy, but this was at the double cost of a premature retirement from the industry and a reclusiveness which itself became notorious. Other stars' 'secrets' quickly became common knowledge. Even William Randolph Hearst, not a star but a producer, was terrified for much of his life that his long-standing affair with the actress Marion Davies would become public (understandably so considering his vituperative denunciations of the 'immorality' of Mae West's films). The star phenomenon is important in considering American society's view of itself, not just because anybody could supposedly become a star given the right combination of luck and talent, but because this illusion of populist democracy was further encouraged by the way in which the stars' lives became part of the public domain. (And not only their printed details either; there was, and I believe still is, a lucrative traffic in guided tours of Hollywood in which the different stars' homes are pointed out to coachloads of rubberneckers.)

Within the films, the on-/off-screen interaction meant that audiences felt (as they still feel) that Rick's distraught drunkenness in *Casablanca* was telling them something about Humphrey Bogart the man as well as about Rick the character or 'Bogie' the star, mediating term between the two. History has intensified this sense of person/character overlap to particular effect in *Citizen Kane*: the grandiloquent but unfinished sprawl of Xanadu has been assimilated over the years to the career of Welles, with its plethora of unrealized or uncompleted projects. There is no better illustration of how Hollywood can commercialize even its rejections or 'failures'.

As in *Casablanca*, so in *Citizen Kane* the political/historical and the personal/biographical are more closely linked than might at first appear. Kane inherits the 'New York Inquirer' and transforms it into a populist pulpit from which slum landlords are denounced and monopoly trusts exposed, but which also invents news with blithe irresponsibility. His 'Declaration of Principles' proclaims: 'I'll provide the people of this city with a daily paper that will tell all the news honestly. I will also provide them with a fighting and tireless champion of their rights as citizens and human beings.'[21] The blend of anti-exploitation crusade and condescending personality cult (present in embryo in Kane's famous declaration 'I think it would be fun to run a newspaper', reprised in what is actually the last shot of the film) is ambiguous enough to provoke several conflicting comments within the film. Thunderous applause greets his election speech, with its promises of protection for 'the underprivileged, the underpaid, and the underfed'; but his friend and counsellor Jedediah Leland delivers a prescient warning: 'Charlie, when your precious underprivileged really get together – oh, boy, that's going to add up to something bigger than your privilege, and I don't know what you'll do.' [22]

His boyhood guardian, the old-school capitalist Thatcher (was it Welles or his scriptwriter Herman Mankiewicz who showed the gift of foresight in choosing that name?), is shown in a simulated newsreel at the beginning of the film denouncing Kane as a Communist; this is immediately followed by another speaker who calls him a Fascist; and Kane then describes himself as 'only one thing – an American'. This parade of identities, laughable through its very rapidity, becomes disturbingly ambiguous in retrospect. Kane is demonstrably neither a Communist (he is far too happy with far too much social and economic power), nor a Fascist (his personal bullying and authoritarianism are not shown as having any ideological correlatives). In one sense very much a stereotype 'American' – in his brashness, his apparent open-handedness, and his man-of-the-people style – he is at a deeper level the reverse of all that, in his moody withdrawal and final crazed retreat into a premonitory incarnation of Howard Hughes. None of the labels proffered in the first moments of the film 'takes', either then or as the narrative unfolds. Kane is as elusive a phenomenon on the ideological plane as he is in his personal life; the 'Rosebud' secret sought after in vain by the journalist Thompson has wider reverberations than the 'dollar-book Freud' referred to by Welles in a famous apology for the film's final moments.

For with Kane's deepest personal secret also, attempts are made to make labels 'take', without success. 'Rosebud' is suggested, with varying degrees of seriousness, as having been a woman he loved, a racehorse he bet on, or simply 'something he couldn't get or something he lost' – one missing piece in a jigsaw too complex to be explained by that alone. The last suggestion, uttered by Thompson who has been pursuing the secret throughout the film, is thereby signposted as the most likely to be accurate; but this is largely undercut by its very vagueness. Kane's 'secret' – public and private – comes infuriatingly close to residing in his unknowability.

Comes close to, but does not. The final narrative sequence of the film is a byword not just in cinema, but in American culture as a whole, and its confusing impact is epitomized in Dwight Macdonald's observation that 'he gets a "big thrill" out of this shot, though he can't explain why'.[23] The shot in question occurs when, after Thompson and the rest of Kane's retinue have given up their quest and the lumber from Xanadu is thrown into the furnace, the camera closes in on a sledge bearing the name 'Rosebud' – the sledge the boy Kane was playing with at home when Thatcher came to take him away for education in his inheritance, the sledge with which he lashes out at his would-be benefactor ... 'Dollar-book Freud', maybe, but dollar-book Freud was to sell very well in the America of the forties, and in the light of what we have said about the importance of the 'loner' theme in cinema of this period the emotional appeal of this resolution is obvious. What is its relevance to the 'public' aspects of the film?

Initially, and rather obviously, it suggests that there is (or was) a Kane who would rather have been left alone, a Kane who as it turned out all but disappeared into the abyss between his private and public life. The presence of the glass ball containing a snow-scene, first in Kane's hand as he utters the word 'Rosebud' and dies, then in the 'flashback' sequence with Susan Alexander in her room (a moment of modest but genuine private happiness), acts as metaphorical reinforcement of this potently sentimental piece of myth-making. But 'Rosebud' not only articulates the fearful gap between public and private, it also works to fill it through its narrative confirmation that it is the masses who always know the 'truth' about Kane. They lionize him as a crusading pressman at the beginning of his career, but reject his political candidacy because of his sexual double-dealing – partly American puritanism, but also a recognition that a promise to root out hypocrisy in public life comes ill from one whose behaviour strongly suggests that his marriage – to a

President's niece – was largely an affair of convenience. The scene in Susan's apartment-block, where Kane roars maniacal defiance at his enemy 'Boss' Jim Gettys, is evidence that his delusions of grandeur, almost of omnipotence, are getting the better of him, a verdict borne out by the absurd lengths to which he goes to promote Susan as an opera-singer. The 'people' reject Kane just when his hitherto sure judgement is beginning to fail him. Their decision is shown to be the right one.

The effect of the final 'Rosebud' close-up is to ensure that this knowledge is carried over from the political to the biographical plane. We see the name on the sledge only after Thompson and his entourage have admitted defeat in their exhaustive research and speculation. The experts do not know; 'we, the people', sitting in the cinema, do. After the close-up of the sledge, the camera tracks away through the grounds of Xanadu, finally focusing on a sign which reads 'PRIVATE – NO TRES-PASSING'. The warning, clearly emblematic of Kane's life, is ironic because now irrelevant. In this respect as in so many others Kane's gigantic enterprise has failed. The film's narrative closure operates an ironic reversal of the ill-defined, but pervasive, populism that has dominated its unrolling under Kane's aegis. Populists should beware of the people.

The Post-War Era

Gregg Toland, who directed the photography for *Citizen Kane*, also worked with William Wyler on *The Best Years of Our Lives*, a lengthy drama about the problems experienced by three ex-servicemen – a soldier, a sailor, and an airman – when they return from war service to their small home-town and begin to readapt themselves to civilian life. Wyler and Welles were both praised by the influential French critic André Bazin for their use of 'deep-focus' – the camera observing and constructing the scenes in depth, within which action could take place on several levels at once – and their preference for long, continuous takes over the sharp editing and juxtapository techniques adopted by directors such as Eisenstein and the German Expressionists.

This predilection, apparently a purely aesthetic one, in fact masks a political choice; Bazin's left-wing social democracy

found an open, fluid visual style within which viewers could select and combine the elements that seemed important to them, preferable to what he saw as the arbitrary, even tyrannical divisiveness of Eisensteinian montage or Expressionist stylization.*
It is no coincidence that the so-called 'Realist' directors Bazin most admired operated within social democracies (usually, as with Welles and Jean Renoir, from a leftish critical standpoint not drastically different to Bazin's own), nor that the two styles he singled out for disapproval emanated from pre-Nazi and Nazi Germany and Bolshevik Russia. The deep-focus of *Citizen Kane* and *The Best Years of Our Lives* was a kind of democracy of the image in action (though it is interesting to note that Bazin's 'democracy' did not extend to giving Toland his full share of credit for this; for Bazin the director as *auteur* was unquestioned king of the films he signed).

In *Kane*, as we have seen, the repercussions of democracy and populism reach out beyond the celluloid narrative to involve the spectators in the questions about knowledge the film poses. *The Best Years of Our Lives* – a much greater commercial success, if only because its theme was so obviously suited to 1946 – provides its audience with a far cosier resolution. Its three servicemen each confront problems in their readaptation to civilian life: the banker approaching middle-age finds the independent development of his family in his absence difficult to accept, the airman who has lost both hands fears to confront his fiancée, and the soldier returns to find the bride he left behind entertaining a string of lovers, and his job in jeopardy. All these problems are resolved in a final reconciliatory wedding-scene in which adequate employment and true love form an unbeatable couple. The life of the small town is not without many features of American small-town nastiness (deceit, gossiping, business hypocrisy), but they are reassuringly overcome at the end. It may well be because its message† is so much more unambiguous than that of *Kane* that the film was so much more successful at the time of its release.

Joan Mellen in her study of the American male in film, *Big*

---

*This is interestingly discussed by Peter Wollen in Chapter Three of his *Signs and Meaning in the Cinema*.

†This term is intentionally used here. Modern criticism has taught us all that books and films are not mere receptacles for messages and ideas, but some have that quality more, or more overtly, than others.

Bad Wolves, draws an interesting parallel between the insecurities of servicemen returning from the Second World War and the depressed state of the South after Unification. This is valid on the interpersonal as well as on the socio-economic plane. The three returning servicemen in *The Best Years of Our Lives*, strangers until brought together on the flight home, quickly feel at ease in one another's company despite the social and cultural differences that precluded their meeting in peacetime. When these reassert themselves in a multitude of ways, 'the ceremony of innocence is drowned' (to quote Yeats's *The Second Coming*), and the apparent irony of the film's title turns out to be less than expected. Wartime is portrayed, for all its maiming horror, as a nonpareil means of bringing men together in brotherly solidarity (while women presumably turn into vegetables at home or animals in night-clubs – their wartime careers play little part in Hollywood movies, with the notable exception of *I Was a Male War Bride*).

This may help to account for the new popularity of the Western immediately after the war, following its slump in the early and middle thirties. Robert Warshow's celebrated essay 'The Westerner' contrasts the worlds of the gangster movie and the Western, the first seen as one of urban romantic tragedy and violent, nihilistic denial, the second as one of rural, classical defeat in which resort to violence in defence of one's honour is tragically inevitable rather than sadistically enjoyable. André Bazin also stresses the epico-tragic element in the genre, in which the sheriff's star is seen as sacramental (like a Christian sacrament, its efficacy is independent of the qualities of its bearer), and adduces the relative absence of close-up shots to show that the Western is concerned with groupings and archetypes rather than with individual character-analysis. Even the nostalgic atemporality of the genre is socially and historically determined, for the 'golden age' it attempts to recreate has much in common with the 'best-years-of-our-lives' view of wartime.

Joan Mellen draws her parallel when discussing Howard Hawks's *Red River* (1948) often considered the outstanding Western of its period and certainly a fruitful terrain for analysis. John Wayne/Tom Dunson leads a thousand-mile cattle-drive through virgin territory; at least, it *would* be virgin if not for the Indians, made responsible for the death of Wayne's fiancée in a short flashback at the beginning. Wayne had left her to embark on his cattle-drive, but her wagon-train was massacred by Indians after his departure. This very short sequence has a great weight to bear in the determination of character in the film. Wayne's ruthlessness, maniacal determination, and eventual lapse into near-psychopathy are all the result of his earlier decision and its aftermath – again relentless public ambition is the product of private loss and grief – and his ambivalent attitude towards Montgomery Clift/Matthew Garth, the young orphan who joins the drive, is explained by Clift's being the son he has never been able to have. Tensions of a family kind thus interlock with the 'professional' ones we should normally expect to find in a cattle-drive Western.

This all sounds very far removed from the world of epic solidarity evoked by Bazin and Warshow, but it is precisely the constant jeopardizing of this world and its fragile reassertion at the end that constitute the social and narrative dynamic of the film. At daybreak, just before the drive moves off, there is a celebrated 360° pan round the wagons waiting to depart. The ambiguity of the lighting and the sense of coiled repose, after the build-up to the departure and before the drama we are sure will follow, both contribute to a sense of quietly ecstatic tension, not unlike the pre-battle sensation evoked in many war films. The peace plus the suspense, the overriding impression of *totality* produced when the camera finally returns to its starting-point, distil in one exceptionally powerful, but still typical, shot the sense of epic male communion so strong in the Western of the post-war years.

The friction between Wayne and Clift oscillates between the personal and the professional, between father versus son and experienced leader versus raw recruit. In the scene where Wayne announces his decision to hang (not merely to shoot) two deserters, Clift's defiant challenge to his authority is predicated on humanitarianism and common sense at once; Wayne is behaving not merely like a barbarian, but like an inefficient and counter-productive one. Thereafter Clift takes over the leadership of the drive, stalked by Wayne at a distance with threats of revenge; the communion magically evoked by the 360° shot has fallen into ragged internecine tension. But it is reimposed, albeit in a socially and sexually more fragile form, by the film's ending. Wayne catches up with Clift at the end of the drive, meets him at the 'appointed place', and engages him in a bitter fist-fight which is interrupted by Joanne Dru (Clift's prospective fiancée, who has proved her 'manhood' in a wagon-train shoot-out) with the cry: 'Whoever thought either one of you would kill the other?' Feuding father and son are reconciled through the 'timeless' intervention of woman, and the film ends with Wayne/Dunson agreeing to add Clift/Garth's initials to his cattle-brand, as he promised to do at the beginning if the youth proved adequate to his task.

So the circle closes, or re-forms; but the quotation marks round 'timeless' in the previous paragraph are not fortuitous. Woman in the world of the Western is almost always a subversive or disruptive force, even if that subversion and disruption take the form of preventing bloodshed. This is tacitly recognized when Wayne leaves his fiancée behind in the flashback at the beginning, and confirmed by the absence of women from his life thereafter, with its implication that celibacy is a precondition for the successful cattleman. (The term 'celibacy', of course, should not in the context of the Western be taken with sacerdotal solemnity, as the abundance of prostitutes in the genre suggests; *emotional* celibacy, exemplified in the dual attitude towards prostitutes as receptacles for sexual relief and companions whose only defect is that they are not men, is what is germane here, though Wayne/Dunson appears to have no sexual relations of any kind.)

The ending of the film, of course, defeats this assumption, not only via Clift's engagement but because there might well have been no Clift to get engaged without the intervention of Joanne Dru (though her remark again casts doubt upon that . . .). The charmed circle of male camaraderie can re-form only through the initiative of a woman. *Red River* is exemplary of the way in which genre conventions can be used subversively as well as stereotypically, and of the complexity of attitudes towards sexuality and friendship in post-war American society.

These are much more directly, because contemporarily, expressed in *I Was a Male War Bride*. Cary Grant as a French army captain meets and marries American army officer Ann Sheridan (after, inevitably, having his arrogant chauvinism disrupted by her pertinacious superiority); he can obtain permission to emigrate to the United States with her only by complying with a set of regulations designed for male officers importing brides from Europe. This involves him in a series of

American Society in and through its Cinema

ludicrous situations with sexual panic never very far below the surface. Turned out of one hostel/hotel after another with the refrain: 'You can't sleep here' (by which title the film is also known), he is eventually forced, in a conscious evocation of his role in *Bringing Up Baby* eleven years before, to dress up as a woman in order to join his bride on the boat. The sexual frustration is double, at his inability to consummate the marriage and at the logical, but intolerable, denial of any sexual identity distilled in the film's title. Peter John Dyer's description of the film as 'a hackneyed and slapdash army farce'[24] is so inappropriate as to lead one to suspect unconscious defence-mechanisms at work; its continued, not to say enhanced, appeal to modern audiences lies in the black play it makes with notions of sexual identity, while in 1949 the bureaucratic confusion it evoked would have been a sufficiently recent memory with many of its audience to provoke rueful laughter. Hard-and-fast definitions, sexual or social, are shown to be potentially threatening as well as ridiculous.

This did not prevent their use to stifle and drive into exile many of Hollywood's finest talents. The House Committee on Un-American Activities (hereinafter HUAC) began its investigation into, and purge of, 'Communists' in the film industry in 1947; but the danger-signals had been visible for a long time, from the adoption of Hays to the domination of the trade-union 'market' by the International Alliance of Theatrical Staff Employees (IA), strongly anti-Communist and in the thirties much given to racketeering. The Hollywood Communist Party was among its own worst enemies, uncritically pro-Soviet, hostile to John Dewey's investigation of the Moscow show-trials, and at times almost as free with its own hate-label, 'Trotskyite', as its enemies were to be with that of 'Communist'. Strikes led by the Communist-backed Conference of Studio Unions in 1945 and 1946 prepared the ground for a confrontation in which the IA, most powerful organ of the so-called Hollywood 'labour movement', was among the most vociferous opponents of 'Communists'' right to work. The statement of IA's leader, Roy Brewer, is worthy of quotation: 'Communists want to use the movies to soften the minds of the world . . . I wouldn't be opposed to their working in an industry where they wouldn't hurt anybody, but I'm absolutely opposed to their working in the movies.'[25]

This is interesting, not only as a further confirmation of the 'uniquely powerful instrument' view of cinema referred to earlier, but as an indication of how fundamentally right-wing in its structure and functioning the Hollywood institution was. An instrumental factor in the choice of Los Angeles at the beginning of the century had been the Californian tradition of hostility to organized labour, and Brewer's statement shows that even such organization as was allowed – not to say encouraged – was willing to defend only those whose ideas were not too far removed from its own.

The epithet repeated *ad nauseam* to damn films and individuals critical of aspects of American society was 'un-American', convenient and revelatory in its vagueness. On the one hand, it supposed the existence of a consensus as to what 'Americanism' was, at once so self-evident and so wide-ranging that failure to conform with it could only be the result of moral and political corruption; on the other, it could be used in self-justifying fashion against anybody whose views the Committee disliked. Ayn Rand's *Screen Guide for Americans*, with its three precepts ('Don't Glorify Failure – Don't Smear Success – Don't Smear Industrialists'), gives a fair idea of what the perception of 'American' was, bewildering though it must have been for even the most dedicated supporter of her views to put into effect (what attitude, after all, was one supposed to adopt towards a failed industrialist?). Individual initiative and enterprise, against the tyrannical narcolepsy of collectivism, was at the heart of the concept, which thus interlocks with the ideology of the market. Milton Friedman has sincerely, if inelegantly, condemned the Hollywood blacklistings as 'an unfree act that destroys freedom because it was a collective arrangement that used coercive means to prevent voluntary exchanges',[26] indicating the division on the American right between the 'libertarians' for whom any restriction on trade is anathema and the 'hard-liners' for whom there is no substitute for fighting collectivism with what are allegedly its own weapons. The judiciary actually found against the hard-liners in 1951, when the Wage Earners' Committee – its members a waiter, a telephonist, a retired salesman, and a restaurateur – were stopped by a court order from picketing films featuring supposed Communists, with such charming slogans as 'Yellow travellers support fellow-travellers'. Whatever the scruples and reservations of the libertarians at the time, there is no doubt that it was the hard-line right that was the dominant element in the McCarthy era.

The inconsistencies and absurdities of their case were legion. A mechanism was set up to help ex-Communist Party members purge their contempt and rehabilitate themselves in the industry, with quasi-Inquisitional zeal – all in the name of the defence of American freedom. The existence of a blacklist was officially denied – yet the lawyer Martin Gang was able to get people removed from it if they were willing to testify to the HUAC. The 'Hollywood Ten' were jailed in 1947 for refusing to answer questions about their membership of the Communist Party, although they had pleaded the supposed constitutional freedom of the Fifth Amendment which permitted them not to answer questions if that might incriminate them. The original HUAC chairman, Parnell Thomas, was jailed for bribery in 1949 and McCarthy himself finished as a discredited, drunken figure. Yet none of this stopped John Wayne, several years after the HUAC hearings, from referring to McCarthy as a 'great American', and denouncing those who were supposed to have hounded him into an early grave. This judgement was a great deal less ridiculous coming from the man widely regarded as the original 'great American' than it would have appeared from almost any other source.

With its stress on virility, independence, and the defence of America against barbarian hordes, the Western was clearly the genre most suited to conveying the values of McCarthyism. Admittedly, *High Noon* (1952) has been interpreted as an allegory of resistance to it, with the lone heroism of Gary Cooper/Kane (was the surname chosen with one eye on Orson Welles?) representing courage in the face of the ethically unacceptable. This reading gained credibility from Wayne's known disapproval of the film's final sequence, where Kane flings his sheriff's star onto the ground in disgust before riding away for ever from the town he has just saved. For the ultimate super-patriot, such blasphemy was enough to destroy *High Noon*'s claims to being a truly great Western (an interesting reflection on the implied hierarchy of movie values which makes the Western the greatest genre, because the most 'truly American').

But other factors undercut the reading and deprive it of conviction. Cooper himself was a 'friendly' witness to HUAC, albeit less ferocious than Wayne. There is no reason to read the townspeople's cowardly evasiveness as compliance with McCarthyism rather than as refusal to acknowledge the supposed 'menace' of Communism in the industry. Grace Kelly's saving of her husband's life by shooting one of his attackers, in breach of her Quaker principles, suggests that ideologies of human brotherhood are simply not adequate to the toughness and ruthlessness of frontier America. Had Wayne looked at the film rather more closely, he would have found much in it to please him.

As it was, he co-starred for Howard Hawks in *Rio Bravo*, made seven years after *High Noon* as a deliberate riposte to it. Wayne/Chance, the sheriff charged with keeping a criminal in jail against the violent efforts of his brother's gang to free him, says of the volunteer posse offered to him: 'If they're really good, I'll take them. If not, I'll just have to take care of them' – a counter to *High Noon* where in Hawks's words 'Gary Cooper ran around trying to get help and noone would give him any'. *Rio Bravo* is an outstanding example of the greater tendency towards professionalism in the Western of the fifties, seen by Will Wright as indicative of the tendency in society and the economy towards large-scale corporatism. The defence of society is seen as a job to be done, as efficiently as possible, by a cooperative unit of professionals, whose means may be unorthodox but fulfil their purpose (as with the use of dynamite at the end of *Rio Bravo*, clearly incongruous with traditional Western iconography but justified by its effectiveness). This is at the opposite pole to *High Noon*, where the hero's motivation is ethical, almost transcendental, his action is carried out alone, and his means – reliance on the combined quickness of his gun and his wits – are ultra-traditional.

The Hawks Western perceives women as a threat to the efficiency of the male group; Dean Martin/Dude in *Rio Bravo* and Robert Mitchum/J. P. Harrah in its remake *El Dorado* both take spectacularly to drink because a woman has betrayed them, and in both films the paternal presence of John Wayne is required to 'wean' them from the bottle back into combative action. *High Noon* is only superficially less misogynistic, for although Grace Kelly saves her husband's life it is at the cost of her highest religious principles, and she is portrayed as yielding to an impulse all too human rather than as realizing the limitations of absolute pacifism (she sobs as she fires the shot and has to be consoled by her husband afterwards). There is no getting away from the anti-woman bias of the Western, nor its cognate yearning for the 'adolescence' of American society when (as in early biological adolescence) true – read homoerotic – companionship among males was supposed to have reigned and women were at best functional, at worst a treacherous nuisance. Wayne's own fulminations against 'faggots' are a defensive recognition of this situation rather than a disproof of it.

No other genre carried distrust of the female so far, for no other genre enjoyed such privileged – though as we have seen only apparent – insulation from the social and historical determinants of its time. The war had given American women a taste of an independence they were loath to renounce in peacetime, but the need to guarantee jobs for returning soldiers came into conflict with this, and according to Marjorie Rosen more women resigned or were dismissed from their jobs between VJ-Day and 1947 than during the Depression. Fear of the 'new woman', no longer submissive and pliable; anxiety about the threat to virility implied by a wife's unwillingness to live off her husband's earnings; nostalgia for the easy male camaraderie of wartime – all played their part in the reaction, and it is not implausible to detect their influence in the number of fifties movies that dealt, on the one hand with occupations reserved specifically for women (notably that of female film-star), and on the other with increasingly ambiguous areas of male sexuality.

1950 saw the two classic 'Hollywood-on-Hollywood' movies, *Sunset Boulevard* and *All About Eve*. Both are clearly pervaded by the industry's obsession (which spread to the whole of Californian culture) with ageing, seen as inevitably a greater threat to women than to men. Anne Baxter/Eve and Bette Davis/Margo Channing engage in what is virtually a set-piece pitched battle, between the outwardly adulatory but in fact ruthlessly ambitious fan and the outwardly vituperative but in fact insecure and unhappy star she supplants in the public's affection. The spoils are evenly divided at the end: Eve wins the Sarah Siddons Award but is deserted by friends and lovers for her hardness (and memorably denounced by George Sanders as the theatre critic Addison de Witt), while Margo finds compensation for her waning career in her love for Bill. Both women get what they want, but the implication is that Eve's career success is a poor consolation prize for the Genuine Emotional

Relationship(s) her ambition has rendered her constitutionally incapable of having. A successful acting career and success in personal relationships are presented as opposites between which a choice has to be made, a perception reinforced by Davis's own string of unhappy marriages, declining career (briefly revived by this film), and assertions that personal happiness was what she most wanted. The film is often read as a condemnation of what Hollywood does to its denizens, but this is only partially true. It is capable of providing every kind of satisfaction they want, for Eve's ambition and de Witt's cynicism but also for Margo's emotional fulfilment (Hollywood-based since Bill is a producer). The neatness of the narrative closure reminds us that Hollywood's larger-than-life quality extends to its omniscient perception of what Eve and Margo want and deserve.

Sunset Boulevard carries even further than *All About Eve* the interplay between on- and off-screen lives of its principal actors. Gloria Swanson/Norma Desmond's butler, in her gloomy morgue implausibly isolated in the middle of Hollywood's most fashionable residential area, is Erich von Stroheim/Max von Mayerling, her ex-director in real life as well as in the film (where his humiliation is intensified by his being also her ex-husband). A technique sometimes used by realist novelists to lend plausibility to their narratives is to set historical and fictional characters side by side (as, for example, Tolstoy does with Napoleon in *War and Peace*). The cinema's historical past is too short to permit it many examples of the same effect; the most famous occurs in *Sunset Boulevard*, where 'Norma' gathers around her for a macabre card-evening a gallery of silent-era celebrities including the gossip-columnist Louella Parsons and Buster Keaton, both playing themselves. The effect of this is complex. It invites us even more than the rest of the film (which includes a clip from Swanson's last great silent success, the Stroheim-directed *Queen Kelly*) to forget 'Norma' and think only of Swanson. Like Bette Davis's misery and humiliation in *All About Eve*, the figures round the table have an ambivalent effect, at once denouncing the Hollywood that has so reduced them and paradoxically glorifying it. For 'there were giants on the earth in those days'; the whole film is constructed round the notion of *scale*, epitomized in Swanson's famous rejoinder to William Holden when he says that she used to be a big star: 'I *am* big. It's the pictures that have got small.'

One of the main reasons why many silent stars were ruined by the coming of sound was precisely their inability to adapt to the different scale it demanded, and it can be argued that the whole history of screen acting has been one of scaling-down, from the grandiloquent gestures initially often imported wholesale from theatre through to the low-key naturalism that has been the dominant mode from James Dean through to John Travolta. The exceptions – Bette Davis, Elizabeth Taylor, Peter Lorre – have distinguished themselves precisely *as* exceptions to a rule. Swanson's descent of her mansion staircase towards the policemen who wait to arrest her for the murder of Holden resumes the distance between the bigness of silent-era acting and the time at which the film was made. So deranged has her withdrawal into her dreams of grandeur become that she can be persuaded to walk down the stairs only by Stroheim's setting up a counterfeit 'comeback' for her, pretending to start lights and cameras rolling as she moves ecstatically towards them. Most audiences would probably not feel at this point that Swanson is 'big' and the pictures have got 'small'; rather, the pictures (1950 version) appear about the right scale, and Swanson's grandiloquence is a sad comment on a style of acting (and by extension a number of actors) which, having failed to adapt to changing conditions, was pitilessly abandoned by the Darwinism of the Hollywood market.

In both these films, then, Hollywood appears double – as cynical tormenter but also as ceaselessly capable of responding to, and helping to produce, its public's desire for novelty. Neither movie, despite the humiliations both inflict en route, can be described as truly tragic, for Swanson as much as Davis is more fulfilled by the end of her film than at any time during its action. Analysis of both has tended to concentrate on the image they produce of Hollywood as carnivorous monster, but it is also shown as a dream-factory able to fulfil the fantasies and ambitions of its own hardest workers. In the love-hate relationship between Hollywood and the American public, love – unsurprisingly since the relationship was after all mediated by Hollywood – nearly always comes out on top.

This is also true of *Singin' in the Rain*, best-known of all MGM and probably of all screen musicals. So famous in its own right has the title sequence become that its narrative role is sometimes neglected; Gene Kelly bursts into song in the film because he (with help from Debbie Reynolds and Donald O'Connor) has just discovered, or invented, song in film. The musical genre, particularly its MGM/Arthur Freed products, has always been associated with lavishness and opulence, wide screens, vivid colours, and a spectacular guying of its own conventions – all characteristics which combine with the song-and-dance numbers to defuse much of the tension, even malice, potentially present in many of the plots. The intrigue of *Singin' in the Rain* is as fraught with malevolent backstage tension as that of *All About Eve*, right up to the climatic moment when Gene Kelly lifts the curtain behind his 'partner' Jean Hagen to reveal that her strident voice (hitherto concealed by the silent movie) is in fact being dubbed by the unknown Debbie Reynolds. When Hagen rushes from the stage in combined embarrassment and fury, it is symptomatic of the conventions by which we read the genre that we do not ask ourselves where she will go. She has been rendered cruelly anachronistic by the film we have been watching, both in its narrative (which pre-emptively destroys any audience sympathy for her by presenting her from the beginning as a shrill-voiced shrew) and in its very performance as a musical. Even a token narrative concession to her at this stage would hardly be enough to atone for the way in which the entire thrust of the film's spectacle has denounced her obolescence.

Serious critical writing on the musical has tended to concentrate on the dialectic or interplay between the flow of the narrative and its punctuation by the set-piece numbers. This became much more fluid in the musical of the fifties, so that it is often as though the tensions produced by the narrative organically generate their own resolution in the shape of a song-and-dance number. A good example is the 'Good Morning' number in *Singin' in the Rain*, where the Kelly/Reynolds/O'Connor triumvirate who have been up all night working out how to turn their silent movie into a musical discharge their combined tiredness and elation in a complicated routine around the apartment. This helps to account for the appeal of the musical to a society of affluence and 'conspicuous consumption'; given a sufficient supply of colour, décor, and musical talent, there was no tension that could not pleasurably resolve itself. Michael Wood has spoken of Gene Kelly's dynamic brashness as symptomatic of the success orientation of

American society in the fifties.[27] Against it, Jean Hagen, as sympathetically contemporary as a pterodactyl, stood no chance whatever.

Thus, even the affluent benignity of the decade's finest musical shows a certain edgy malice in its evocation of its own milieu – and, by extension, that of American society. An attentive reading of *Funny Face* will likewise bring out anxieties and hostilities behind the fluency of colour, narration, and performance, sacrilegious though it may be thought even to make such a suggestion about a film starring Fred Astaire. The adjective recurrently used to describe Astaire is 'timeless', understood both biographically and historically. In a society obsessed with the process of ageing, he only occasionally, and then always with perfect grooming, hints at approaching middle age; in a society in which from about the date of *Funny Face* (1957) fashions changed with bewildering speed, his habitual uniform of white tie and tails (varied by expensively classic casual wear) put him above such vicissitudes.

Funny Face can best be understood as a twofold lampoon on fashion – sartorial and intellectual. The authoritarian fashion editress speaks at the beginning of 'the Great American Woman who stands out there naked waiting for me to tell her what to wear', and derides Astaire's choice of Audrey Hepburn (found working in a Greenwich Village bookshop) as the season's new model. The trio fly to Paris, ostensibly to model the new fashions; but Hepburn is more interested in the latest 'philosophical' tendency, Empatheticalism (!). The 'Bonjour Paris' sequence when they arrive is an interesting reflection of American views of the city. They are shown separately, wandering through different parts of Paris and singing lyrics which indicate their different reasons for being there. The editress walks down the Rue de la Paix, centre of the fashion district, singing of trends and designers; Astaire is a predictably convincing boulevardier strolling down the Champs-Elysées; and Hepburn's beat is the intellectual and artistic areas, where she audaciously but inappositely rhymes 'Sartre' with 'Monmartre' (the existentialists' headquarters were in fact in St-Germain-des-Prés and Montparnasse, on the Left Bank). The 'Empatheticalists' turn out to be a tawdry array of charlatans, their leader's advances to Hepburn being more appropriate to the Hollywood 'casting couch', and it is left to the invincible combined forces of Astaire and true love to rescue her and return her safely to New York. The film's opening number, 'Think Pink' (clearly not meant to be construed politically), displays a brash and dictatorial attitude towards clothing fashion less hypocritical, but no less inappropriate, than the Empatheticalists' attitude towards philosophical fashion. Astaire's 'timelessness', in dress and behaviour, is set against these parodied extremes. Mistrust of Continental intellectuals and entrepreneurial women contrasts with the durable delicacy of America's most reliably charming male, and a lucrative career as a fashion model (an area where female and male were not competing for the same jobs) is shown as in every sense more rewarding than Hepburn's ill-advised sally into the universe of concepts.

So much, then, for Hollywood's concentration on specifically female areas of employment. The other area in which sexual insecurity was ventilated in the cinema of the fifties, that of male androgyny and homoeroticism, has already been discussed in the Western. Montgomery Clift, co-star of *Red River* (in which his is indisputably the finest performance), and James Dean in his three feature films (*Rebel Without a Cause*, *East of Eden*, and *Giant*), both created tense and rebellious personas that seem to have had much in common with their tormented, largely homosexual, and brief lives off-screen. *Rebel Without a Cause* is of particular interest as probably the first film to have suggested that juvenile delinquency was not the result of material deprivation alone. His family in that film is comfortably-off and makes sporadic, though inadequate, attempts to 'understand' him. The film proffers an implicit explanation for Dean's rebelliousness (and for his homoerotic interest in the young Sal Mineo) when his father (Jim Backus) appears in an apron, helping to do the dishes; a crisis of manhood, it is suggested, is likely to be quasi-hereditary as well as environmental.

Masculinity, then, was often shown as problematic in the cinema of this time, and this often went hand in hand with a fear of women that verged on the grotesque. Photography of the female anatomy has always tended to be much more fetishistic than that of the male, focusing on different zones of the body at different periods; garters and tight sweaters were both banned by a censorial edict in 1944, so there was an immediate upsurge in close-ups of the female posterior. The fifties witnessed a comparable obsession with large breasts, a tune played quietly and with relative subtlety by Marilyn Monroe before its grotesque fortissimo orchestration by her successors, notably Jayne Mansfield. The very title of *The Girl Can't Help It*, and the string of gags that the film produces around Mansfield's breasts (notably one where she passes a milkman holding a bottle of milk which promptly boils over . . .), indicate the balance between freak-show amusement at her proportions and panic at their presumed implications of sexual aggression and voracity. 'Momism' is a term often used to mock the American preoccupation with and sentimental devotion to the mother-figure (already seen in *The Public Enemy*). I am tempted to copyright 'mammism' as the term for its (in every sense) deformed offshoot, conjoining a covert desire to return to the mammary source of nutrition with a defusing of female sexuality through ridicule that speaks volumes about the male insecurities of those years.

The Sixties

'Momism' is one of a number of interconnected American targets flayed by Alfred Hitchcock in *Psycho*. Hitchcock had been away from Britain for twenty years before he made this film in 1960, and it is easy to forget that he was still very much an expatriate observer of the American scene (as the carefully nurtured 'Englishness' of his accent and persona remind us). Anthony Perkins/Norman Bates's devotion to his dead mother is so pathologically intense that it extends not just to necrophilia (he keeps her mummified body in the cellar) but to an incestuous violation of her biography. Raymond Durgnat says of the film's narrative resolution:

> We expect the clichés: poor mixed-up kid, it was all the fault of stern, possessive, puritanical Mom. But gradually we realise he's [the police psychologist] not saying this at all. It was Norman who was jealous, who imagined that his (for all we know) normal Mom was a promiscuous Mom and murdered and embalmed her and then imagined she was a jealous puritanical Mom and then lived out two false characters – nice normal Norman and nasty Mom.[28]

'Momism', in other words, is distorted beyond its normal field of excessive sugary devotion (of which Elvis Presley, drafted into the army the previous year, provided a textbook example during his period in uniform).* Norman's relationship to his mother effectively fuses the two sides of the Oedipus complex – love of the mother, jealousy of and urge to kill the father – into one. Perhaps this grotesque conflation, read in conjunction with the intolerable sapience of the psychologist, is a mockery of the preoccupation with popularized psychoanalysis and 'psychological explanation' which we have seen was typical of post-war America?

It has been noted earlier that a dislike of the city and a desire to escape to the countryside were prominent features of many post-Depression films. In *Psycho*, the city is presented to us at the beginning as a sweaty, claustrophobic world, full of crudity and duplicity, and obsessive about precise time-keeping (the film's first shot gives us exact time and date, and Marion has to leave the hotel of her assignation in order to return to her office after the lunch-break). But the countryside, all swamps, rainstorms, Gothic mansions, and homicidal horror, is hardly an improvement; and it is doubtful whether the city's chronometric neurosis is any worse than the horrific manner in which time has stopped in the country. Similarly, the clammy swamp in which Norman immerses the cars of 'Mom's' victims may be more unpleasant, but is less threatening, than the 'purifying' shower in which Marion, relieved by her decision to return the money she has stolen, is knifed to death by Norman/Mom.

The publicity surrounding the film at the time of its release made great play with the shower murder, which drew down predictable (and, I suspect, desired) opprobrium upon itself. But the scene is no more unpleasant than the rest of the film, whose onslaught upon the togetherness of the American family and the inbred backwardness of the rural 'simple life' has also a consistent, and undeniable, misogynistic thrust (literally so in the shower scene). The fascination the film has exercised almost uninterruptedly from the time of its release, over audiences ranging from commercial-circuit horror fans to college film societies, undoubtedly resides in its thoroughgoing anti-Americanism.

Psycho shows how, despite the notional influence of Hays and

---

*This is not meant to suggest that Presley's devotion to his mother was either pathological or insincere, merely that – with conscious or unconscious insight – the publicity apparatus around him saw how useful in maintaining him in the public eye its cultivation could be.

his cohorts, the American cinema had moved steadily in the fifties towards greater sexual and emotional explicitness. The Kinsey Report in 1953 startled a whole culture into realization of how it had repressed awareness of its own sexuality. Teenagers acquired greater economic independence with affluence (as in *Rebel Without a Cause*), and the booming rock-and-roll industry (satirized in *The Girl Can't Help It*) quickly became an important wealth-producing sector of American society. Teenage movies (*The Blackboard Jungle* and the James Dean and Elvis Presley vehicles being among the first examples) emerged as a separate marketable category, followed at relatively little distance by the nostalgia industry; *American Graffiti* was already able in 1973 to recommercialize the sounds and styles of only eleven years before.

The nostalgia industry can best be understood for our purposes as an important species of secondary commercialization, a phenomenon whose incidence casts helpful light on a society's different views of itself. Eleven years was enough, in the early seventies, for an earlier teenage way of life to be evoked again as a quasi-anthropological curiosity for those too young to have experienced it. The converse has also often been true; Alan G. Barbour says, rather patronizingly, that 'in an age where the iconoclast seems to be the rule rather than the exception, Bogart appears to stand as the revered founder of the breed'.[29] The college and university film-society circuit rediscovered Bogart in the sixties largely because his interplay of idealism and cynicism chimed well with the guilty affluence and social conscience of the period.

Sometimes, this form of secondary commercialization had a quite explicit political thrust, as when the Greta Garbo comedy *Ninotchka*, about a solemn-faced Soviet agent seduced into uncharacteristic laughter by the capitalist West, was re-released in 1947 at the behest of the Motion Picture Alliance for the Preservation of American Ideals (a forerunner and front organization for McCarthy). Sometimes, it takes the form of a radical rereading of work hitherto dismissed as commercial pulp, as has happened with many 'women's pictures' of the forties and fifties, notably those signed by Douglas Sirk. Sirk, a Dane who began working in Hollywood in 1942, directed thirty films there in seventeen years, mostly low-budget melodramas for Universal, whose enforced reliance on stylized sets and décor and extravagant implausibility of plot condemned them to critical neglect or patronization until well after the end of his career. In the early seventies, his *oeuvre* was rediscovered and became the focus of widespread attention in the United States, Britain, and Germany, for a variety of reasons. There was the interest of the prolific and voluble German 'New Wave' director and critic, Rainer Werner Fassbinder, in the possibilities for distanced analysis Sirk's stylization opened up; there was the increased intellectual preoccupation, forming a virtual academic sub-industry, with manifestations of so-called 'popular' culture and what they could tell us about the society that had produced them; and there was the important influence of feminism and Freudian psychoanalysis (not 'dollar-book' so much as 'five-dollar-magazine' . . .) upon cultural, and particularly cinematic, criticism.

This is important because it illustrates the complexity of ways in which a body of films such as Sirk's can tell us about American society. His work used well-known but not superstar actresses and actors (Rock Hudson, Dorothy Malone, and Robert Stack figure on several occasions) and enjoyed commercial as well as critical success, but consecutively rather than contemporaneously; it is easier in London nowadays to read about a Sirk movie than to see one. The two types of commercialization are both based on Sirk's passionate yet critical anatomy of the American family, to which his fifties audiences belonged and in the denunciation of which his seventies and eighties audience (myself included) had an interest. The typicality of his families often resides in their extraordinariness; the aviators in *The Tarnished Angels* live more precariously, the oil-family in *Written on the Wind* more extravagantly (in both a financial and an emotional sense), Jane Wyman and Rock Hudson in *All That Heaven Allows* closer to the love/hate bond between town and country, than any audience would reasonably have accepted on the plane of 'realism'. It is this 'larger-than-life' quality in his dramatization that gives Sirk's films their dual appeal.

The trend towards 'professionalism' already detected in the Western was also, and for much the same reasons, apparent in the fifties and sixties gangster movie. Labour-movement racketeering, as we have seen, widespread after the end of Prohibition, is at the centre of *On the Waterfront*, a film like *High*

Noon highly ambiguous in its ideological resonance. If Marlon Brando's heroic stagger to work in defiance of brutal intimidation can be seen as an exhortation to clean up a trade-union movement whose corruption was legendary, we can also read it as an invitation to the 'honest working man' to align himself with his employers, if only as the lesser of two evils. (History outside the film again affects my reading of it: its director Elia Kazan, be it not forgotten, had named names to HUAC in 1952, after a brief earlier involvement with the Communist Party.)

What is suggested in *On the Waterfront*, and explored in greater depth in other gangster films of the period, is that the competitive organization of American society has extended so far that employers and unions (or gangsters and police, or Communist spies and those assigned to catch them) have become organizationally almost indistinguishable. It is an attested fact that the wealth and power in the more corrupt unions of the time made them as lucrative a career-structure for the personably unscrupulous as the Prohibition racket had been for Cagney's *Public Enemy*. The union leaders in *On the Waterfront* deploy the full sordid armoury of goons and hit-men; between employers, gangsters, and trade unionists, the difference is at once well-hidden and so slight as to be almost non-existent. This receives a more subversive inflection in *Underworld USA*, made seven years later. Here the world of gangsterdom is assimilated to that of big business (as the film's title, with its 'limited company' overtones, implies).

The concept of revenge in the Western, as we have already seen, is a good deal less problematic. Ringo in *Stagecoach* transforms his mission of personal vengeance into one of social purification; Kane in *High Noon* ironically reverses this, for his killing of the Miller brothers and departure from Hadleyville is a revenge on the cowardly selfishness of the community as well as an act of salvation. But the Western – a much more epic genre than *film noir* within which Fuller worked – has far more archetypally clear-cut ethical divisions. *Underworld USA* is also marked by having been made in the aftermath of the McCarthy era – a period during which ideological and personal rivalries within Hollywood became blurred, and in which last year's friend could easily become this year's denouncer. Its ambivalent position, straddling cynical self-reliance and the hope – articulated but not explicitly realized – of a better, cleaner United

States, thus speaks eloquently of the time at which it was made.

For Fuller's – and possibly Hollywood's – most cauterizing denunciation of how American society warps and deforms the individual aspirations it nurtures, we must look to *Shock Corridor*. The protagonist, Johnny Barrett, feigns madness so as to be admitted to a mental home in which a mysterious murder has been committed. By infiltrating the asylum society and uncovering the killer's identity, he hopes to win American journalism's most coveted award – the Pulitzer Prize. Through painstaking empathy with the three witnesses to the crime, all of whom he manages to coax into brief moments of lucidity, he identifies and denounces the assailant – but at the cost of his own sanity and a relapse into catatonia at the end (perhaps an echo of Norman Bates in *Psycho*?)

The film thus works as an ironic parable on the allied dangers of ambition and knowledge (the knowledge Johnny seeks so desperately, of the murderer's identity, eventually costs him all other knowledge, including that of his own identity). This is borne out by the character of the former nuclear scientist Boden, driven mad by the burden of what he knows and the responsibility he has accepted in pursuance of his career. But the other two inmates whose worlds Johnny penetrates provide a rather different picture. The Negro Stuart has been driven by the racial hatred that greeted his attempt to enrol at a Southern university into dressing up in Ku-Klux-Klan clothing and screaming racist slogans. Trent, the ex-soldier, has been brainwashed by the Communists in Korea, and it is the realization of the error of his ways, and the rejection of his attempts to recant, that have led him to insanity. Trent is thus often seen as an illustration of Fuller's supposedly simplistic anti-Communism, but his case is in fact more complex. He was an easy candidate for brainwashing because of his lack of education, the 'cabbage' with which his head was filled, and the absence of any knowledge of his country's achievements. Stuart and Trent thus both indict American educational failure, as Boden and Johnny indict the other side of the coin, the obsessive drive towards success. That all four are mad by the end of the film goes to show how vehement an attack on the American misuse of individualism it is.

At stake in both *Underworld USA* and *Shock Corridor*, in very different ways, is the destiny of the individual confronted with various forms of what was described in the introduction to this section as 'the organization'. The underworld as big business in *Underworld USA* is in some senses paralleled by the various organizations in *Shock Corridor*; that of the asylum itself, that of the worlds of journalism and education (both connected with the production of knowledge), that of military power, and, via Johnny's sexually paranoid hallucinations about his stripper girlfriend Cathy, that of sex as commodity. If I speak of the 'destiny of the individual', it is not that I think such a phrase remotely adequate to describe what happens either to cinematic characters or to real-life human beings, but rather because conflicts and tensions within American society are usually translated in its cinema into a battle between organization and isolated individual. One form this often takes – Tolly's itinerary of revenge in *Underworld USA* and Johnny's journey through (or rather *into*) the labyrinth of the asylum are both examples – is that of the odyssey, the individual quest involving a voyage through society.

It was Hitchcock who gave this its most daringly literal formulation, in *North by North-West*. Cary Grant/Roger O. Thornhill, a memorably feckless businessman, is mistaken for a CIA agent (in an ironic reversal of the paranoia of the McCarthy era), and pursued by a gang of urbanely vicious Communists. The film has received a multiplicity of readings: Oedipal (Thornhill is shown as comically subjected to his mother at the beginning), geographical (the film's title is a non-existent compass point, and Thornhill's journey takes him through much of the United States, from the United Nations Building in New York to the vast busts of American Presidents on Mount Rushmore), ethical (Thornhill moves from Madison Avenue glibness to a greater personal sensitivity and responsibility, via his relationship with Eva-Maria Saint/Eve Kendall), and comical (witness the practical joke of the non-existent compass point, and Hitchock's own appearance missing a New York bus – missing the journey of his own film).

All these readings overlap and converge at the point of the odyssey. Ideologically, Thornhill represents the brash but lovable forces of American individualism, against the chilling collectivism of his Communist opponents – a variant of the individual-versus-organization theme. Emotionally and psychically, he journeys away from classic American dependence on the mother-figure to a more equal relationship. Geographically, as Raymond Durgnat points out, '. . . the shrines of idealism (the UN Building, the Mount Rushmore Presidents) are also the locales of extreme violence'.[30] The stock iconography of American democracy is subversively deployed, *its* organization too called into question. The film's humour derives largely from the discomfiture of one set of values, founded on fast talking and all-purpose cynicism, at the hands of another, founded on cooperation and the willingness to take genuine risks with one's own life rather than surrogate ones with other people's money. The individualism of the marketplace and Madison Avenue, in other words, gives place to an individuality which it is possible to locate, if at all, somewhere between Hitchcock's practical joking and his Catholicism . . . But that is another question: what is important for us is the multiplicity of ways in which *North by North-West*, a literal journey encompassing a number of metaphorical ones, calls American self-valuation into question while still leaving its individual basis largely intact.

The concept of the odyssey helps to link a genre that had passed its peak by the sixties with one that came to the fore at the end of the decade – respectively, the Western and the road-movie. This is not to suggest that genres simply 'take over' one from another like runners in a relay race, without contradiction or overlap; Westerns are still being profitably made and *It Happened One Night* (1934) was already a kind of road-movie. But the epic-heroic vision of the classic types of Western became increasingly difficult to sustain from the late fifties onwards. John Wayne/Ethan Edwards in *The Searchers** achieves a kind of personal heroism by rescuing Debbie; but his departure at the end is very different to the narrative resolutions of other Wayne classics. Ringo in *Stagecoach* leaves the 'wilderness' that is one pole of the classic Western duality for the 'garden' of happiness with Dallas in Mexico. The joining of Tom Dunson's and Matthew Garth's cattle-brands at the end of *Red River* prefigures a future of prosperous collaboration (or, more elegiacally, Dunson's readiness to bequeath his empire to a worthy successor). But Ethan Edwards rides away once more, not because there are new realms for him to conquer, new areas in which his

---

*For a suggestion of how detailed analysis of this film could be undertaken see the introduction.

individuality can make itself felt (as with Errol Flynn at the end of *Dodge City*), but because he simply does not belong in the world of the garden. His mission as reconciler of garden and wilderness finished, there is nothing left for him but to disappear. Wayne's departure is moving not only because it rhymes with his arrival at the beginning, but because a Western where at the end John Wayne can find no place for himself appears as a kind of elegy for the genre, a muted warning that the old epic certainties are subject to the destructive action of entropy.

The reasons for this are not far to seek. The litany of American disasters and discomfitures over the past thirty years – McCarthy, Korea, the 'youth rebellion', the Kennedy assassination, Vietnam, Watergate – is drearily familiar. Space does not permit a full analysis of how each of these percolated through into the national cinema, though the depoliticization of Vietnam in the late seventies will be dealt with in some detail; but it is obvious that a form so anachronistic from the beginning as the Western – the last bastion of the epic – could hardly survive the massive haemorrhage of national self-confidence intact. Retreat into the camp nostalgia of self-parody (the *Dollars* series), ironic reversal (*Little Big Man*), and transposition of its intrigues, stars, and conventions to a city setting (*Coogan's Bluff*), were some of the strategies adopted, but to apply the term 'Western' to three such different types of film as those quoted is to show how tenuous the homogeneity of the genre by this time had become.

The Western had pitted the individual against a (usually corrupt) organization, as with Kane and the Miller brothers in *High Noon*; it had also shown how a youthful capitalist, even pre-capitalist, society could use cooperation along with competition as a means to organize itself effectively along market lines (for example, the voluntary amalgamation between Dunson and Garth in *Red River*, and the professionalism evidenced by John Wayne's sheriff in *Rio Bravo* when he says of his posse: 'If they're really good, I'll take them. If not, I'll just have to take care of them'). The last example shows how competitiveness could, on the screen as in post-New-Deal society, become mollified by a kind of social conscience which itself paid economic dividends. But beyond this point the genre could stretch only with difficulty.

The connection between the rise of the road-movie and the

insecurity of American society in the sixties and seventies seems at first as apparent as that between the self-confidence of mercantile capitalism and the popularity of the Western. But, if it is true that the ideological façade of individual competition can periodically crack to reveal areas of contradiction and insecurity, we can logically expect to find a similar phenomenon at work in reverse in films that appear to challenge the hitherto dominant values of American society. The obsessive need to travel articulated in the fifties by 'Beat' writers such as Kerouac and Ginsberg (to say nothing of Brando's biker in *The Wild One*); the revulsion from a culture of acquisition and investment and the attraction to one of hedonistic extravagance, via sex, drugs, and drink; the zoological fascination with the bizarre, the financially unproductive, those whom John Wayne and the Daughters of the American Revolution would have joined voices in condemning as 'perverted' – all these phenomena have a social charge so self-evidently 'un-American' as to make any deeper questioning of their resonance appear almost impertinent.

But to subvert is not the same thing as to confront. Much of the most overtly political film-making of the sixties and seventies went on in areas far removed from the commercial cinema – through the work of activist groups, 'underground' film-makers, and newsreel cooperatives. This was only to be expected, but it is worthwhile looking at one or two examples of the quasi-'underground' fringe of Hollywood in order to see how far their critical reflections on American society go, and – more important – where their limitations lie.

Bonnie and Clyde constructed its bank-robbing central characters as sympathetic outsiders, treated Warren Beatty/Clyde's impotence with rather more delicacy than might have been expected, and turned a liberal spotlight on the influence of the Depression on crime. But it is neither so original nor so progressive an intervention in the mainstream of American cinema as this might suggest. *You Only Live Once* (1937) and *They Live by Night* (1948) are far more penetrating in their analysis; impotence for the real-life Clyde Darrow was only half the story, for he was homosexual; and the allusions to the Depression (such as clips from *Gold Diggers of 1933*) work primarily as exotic period colouring. The exhilaration of the robbery- and chase-sequences does help to suggest how crime could have been important as a source of excitement as well as of income in the grey Depression years; but in the context of 1967, when it was made, it smacks rather of another down-market *frisson* for sensation-seeking swingers. There is at least a case to be made out for considering it one of Hollywood's most offensive examples of radical chic.

Repression of homosexuality (or at least homoeroticism) runs consistently through the road-movies of this period. While there is never any suggestion that either Peter Fonda or Dennis Hopper is homosexual in *Easy Rider*, they appear a good deal happier with their motorcycles and each other than with the women they pursue and dream of. Here as in *Bonnie and Clyde*, the theme of the individual breaking away from or disrupting social organization meshes uneasily with that of the market. Bonnie and Clyde are thieves not only in order to live well, but also to provide themselves with a profession, a counter in the social game (thus Bonnie's repetition of 'We rob banks'). The two motorcyclists in *Easy Rider* frame their odyssey, at first sight an impeccably unconventional rejection of the values of bourgeois America, with market transactions of the most dubious kind. Their voyage is funded by the immensely profitable sale of a large amount of cocaine; then, hiding the money in the carburettors of their motorcycles (on which the Stars and Stripes is significantly emblazoned), they set off. Nothing could be more classically mercantile-American than thus to work hard and sell hard to pay for one's holiday.

Their arrival at the New Orleans Carnival is the counterweight to this – for selling, read buying. They visit the finest brothel in town and gleefully take their pick of the female flesh on offer, though the commercial aspect of the transaction is obfuscated by the effects of LSD (under which their love-making in a deserted cemetery comes to appear spontaneous rather than financially determined). It is easy to 'justify' this, as to excuse other dated aspects of the film like Dennis Hopper's solemn homily on the dangers of alcohol and the mind-opening delights of marijuana, by reference to the time at which it was made, when the psychedelic wave was beginning to die away and feminism was not widely recognized as a serious force. But this is hardly the point. Female flesh, like cocaine, marijuana, even (for Jack Nicholson's likeable degenerate) Bourbon, is a commodity first and last. The 'drop-outs' in *Easy Rider*, like those in *Bonnie and Clyde*, are not so much disgusted with American commercial society as eager to branch out into unorthodox, but lucrative, sectors of it.

Sex as commodity reaches its apogee in the Warhol/Morrissey films of the late sixties and early seventies, such as *Flesh* and *Trash*, where it is just one object among others to be bought and sold with blasé (even gay) abandon. This notion percolated through to the commercial cinema in somewhat dilute form in *Midnight Cowboy*, where the possible homosexual bond (between Dustin Hoffman/Ratso and Jon Voight/Joe) is once again coyly hinted at and heterosexuality is the commodity, but with a difference. Here it is the women who pay – or are supposed to; Joe's butch blond splendour is not without admirers, but is less lucrative than he and Ratso had imagined. Again, the individual's odyssey turns out to be inextricably bound up with market forces, and the complex social organization of the big city (New York) defeats the naïve enthusiasm of the provincial. The three films just analysed (two clearly road-movies, *Midnight Cowboy* a transposition of road-movie intrigue and iconography to the big city) all end in death: *Bonnie and Clyde* in an inevitable shoot-out, *Easy Rider* in the murder of the motorcyclists by rednecks, *Midnight Cowboy* in Ratso's death from tuberculosis as the Greyhound bus finally delivers him and Joe in sunny Florida. Law and order avenge themselves on the criminal, the lumpenprovincial on the sophisticated, and the world of urban grime and pollution on the 'little man' who had dared to try to beat it at its own game. The glamour of a certain style of dropping-out finally reveals itself to be too close to the society against which it is supposedly reacting to be able to survive without it.

The Seventies – Violence and Vietnam

The big-city odyssey began to take a particularly unpleasant form at the turn of the decade, with the emergence of the so-called 'vigilante film'. Important early examples of this are the films directed by Donald Siegel with Clint Eastwood, notably *Coogan's Bluff* and *Dirty Harry*. Eastwood plays a sheriff in both of these, but one a long way removed from the traditional iconography of the Western. In *Dirty Harry* he abandons his badge in disgust at the restraints placed upon him by the need to conform to the law. David Thomson says: '. . . that wry, bewildered conservatism has been seen elsewhere at the Watergate hearings of law-enforcement men led into burglarisation'.[31] As an index of the cynicism that pervaded American society between 1968 Berkeley and 1972 Watergate, the Siegel/Eastwood films are ominously revelatory.

The brash humour of *Coogan's Bluff*, in which Eastwood/ Coogan's 'innocent' sexual aggression recalls that of Don Murray/Beau in *Bus Stop*, draws attention to rather than disguising the film's more unpalatable aspects. Nowhere in Hollywood are women more comprehensively reduced to the status of objects than in the vigilante movie, and Coogan's hick naïveté is quite unperturbed by the fact that his gross attentions to the social worker Julie have disastrous consequences for her client. The villain is a representative of the hippy counter-culture (normally associated with non-violence) against which the vigilante genre is a reaction; to quote Alan Lovell: '. . . in the film, challenges in terms of dress, sexual attitudes and drug-taking are converted into a familiar threat of violence – see the scene in the pool-room when Coogan is beaten up by a gang who belong to an older tradition of villains in the American cinema, pool-room hoods'.[32]

To describe *Coogan's Bluff* as a vigilante film is only partially accurate, for even while under arrest Coogan remains a card-carrying sheriff throughout, and the film's very title is a pun on the duality of his role. It suggests at once a tactic of duplicity (Coogan as vigilante/detective) and a rocky outcrop of some neo-Fordian landscape (Coogan as Western sheriff), just as the final motorcycle chase through the Cloisters is poised between the urban (the Manhattan locale, the means of transport) and the Western (the type of chase – riding rather than driving – and the locale's comparatively rustic character). Alan Lovell suggests that 'the basic impulse of the film towards ironic comedy drawn from the "cowboy comes to the big Eastern city" plot is confused by a commitment to male physical and sexual dominance'[33] – a reading of the film as an uneasy blend of two sets of genre conventions, in turn reflecting on the uneasy relationship of Western and Eastern America. The 'Western' values Coogan brings to the big city are rather less likeable than those of his predecessor-as-hick Mr Deeds, consisting as they do primarily of stupidity and coarseness. That these finally carry the day, enabling him not only to return in triumph with his prisoner but also to obtain a forgiving farewell from Julie at the airport, is one of the film's most worrying aspects.

We have seen that the city (especially New York) was perceived as a threatening locale in much cinema of the thirties; forty years later, the perception recurs with a new edge of distaste, and (as *Coogan's Bluff* suggests) without the compensatory idealization of the rural found in many films of the New Deal era. *Taxi Driver*, probably the most notorious of vigilante films, raises this distaste to the level of self-engendering paranoia. Robert de Niro/Travis Bickle is like a 'painting-by-numbers' version of the American existential anti-hero. A Vietnam veteran (whence his headaches and insomnia; whence, too, a whole discourse of persecuted self-loathing for which no other episode in American history could so readily have served as a cipher); a night-time taxi-driver in and around the Times Square area of New York, both focus and stimulus for his sexual nausea ('Every night, I have to clean the come off the backseat'); dementedly torn between two women, the WASP whose crystalline purity he seeks to defile by taking her to a pornographic movie and the twelve-year-old prostitute whom he restores to her parents at the end, after eliminating her 'protectors' – even in this sentence it is easy to observe how the language of (attempted) analysis slips into that of sensationalist rhetoric, a rhetoric catalysed by Bickle's case-history but stretching in its repercussions far beyond that.

The obsessions of *Taxi Driver* are those of the post-Vietnam era and the vigilante movie at their most strident, and stretch back in American culture and cinema a good deal further. The city is a cesspit of iniquity so fascinating that there can be no question of tearing oneself away from it. Women are whores or goddesses of purity, and can be conceived of only at these two poles, the mobility between which was graphically demonstrated after Marilyn Monroe's suicide, when salivating desire for the 'sex-symbol' metamorphosed, under the combined influence of guilt and commercialization, into posthumous canonization of the 'wide-eyed innocence' beneath the erotic mask.* Vietnam is

---

* Witness Ken Russell's use of the Monroe pose from *The Seven-Year Itch* as an icon of healing in *Tommy*.

evacuated of any historical specificity to become perilously convenient shorthand for the combined traumas of the sixties and seventies, of which it was but a part (though undoubtedly the most significant). Against a crazed world, only a demented individual can hope to wage successful war. This is the logic of *Shock Corridor*, with whose sexual excess (Johnny's hallucinations, the nymphomaniac ward, Boden's regression to the pre-pubescent) Bickle's Manicheism in many respects rhymes. But now, twelve years later, we are out on the open streets, with Bickle's malady odiously proffered as a latter-day version of chivalry which is all the film has for us to admire.

For there is little doubt that the end of *Taxi Driver* constitutes Bickle as its hero, not only through his elimination of the pimping villains and return of Iris to her adoring parents, but also through the reaction of the WASP character Betsy, played by Cybill Shepherd. We see her at the end admiringly eyeing Bickle as he drives past in his taxi, but his attitude is one of cool detachment. The importance of the look as a means of directing the libidinal – and, by extension, ideological – energy of a film, and of articulating attitudes of and towards women in particular, has received much attention from recent critics. Sex-symbols such as Monroe and Mansfield are obvious focuses for the objectifying male gaze, but in both *Coogan's Bluff* and *Taxi Driver* it is the male who is the object of admiring female attention. Julie looks at Coogan in his helicopter, as Betsy looks at Bickle in his taxi, in such a way as to make it quite clear that both males' earlier grossness and sexism can be forgiven now that they have fulfilled their essential role as heroes and captured or eradicated the villains. This retrospective vindication of their behaviour is given a particularly repellent inflection in *Taxi Driver* through Betsy's being a campaign worker for a presidential candidate clearly designated as being – by American standards at least – on the Left. To the whole complex of liberal attitudes implicit in Betsy's culture, her resentment at being linked with a world of sexual exploitation, and her working with others to bring about social improvement, the film opposes a theology of revenge and a pathologically unrestrained individualism as savage as they are reactionary.

Taxi Driver stands at the antipodes of 1968 American counter-culture – not only ideologically, but also geographically. It is predicated upon New York violence just as surely as 1968 was upon Californian peace and love. The contrast between East and West Coast – often seen more specifically as one between Manhattan and Hollywood – is important in many other American films of this period. That the nostalgic cruising idyll of *American Graffiti* is set in a small town in California, and the large-scale organized violence of *The Godfather* in New York, may serve as a touchstone of how the two coasts and cultures were often constructed in American cinema.

Woody Allen's recent work is a refreshing – and important – exception. In *Annie Hall*, his chagrin at losing Diane Keaton to a Los Angeles record-producer is not just that of a defeated lover; it is that of a Manhattan interloper, smaller, less tanned, less bland, and more sardonically coherent than the denizens of what he dubs 'Munchkin-Land'. In *Manhattan*, it is to London – not California – that he loses the girl (Mariel Hemingway) at the end, and the West Coast intrudes only in the safely neutralized form of celluloid nostalgia. His patient explanation of exactly who Rita Hayworth was, like his embarrassed quoting from *Casablanca* (itself the founding myth for his earlier *Play It Again, Sam*), belongs to a California he knows and – especially with a younger woman – can control, that of the supposed 'golden age of Hollywood'. Joan Didion's much-publicized criticism of Allen's work – *Manhattan* in particular – for the alleged narcissistic superficiality of its cultural allusions is ironic, if only because the target of much of Allen's satire is precisely that narcissism, in both its cerebral-Manhattan and its hedonistic-Los Angeles manifestations.

One aspect of American society overridingly rejected by its sixties and seventies cinema is the stress on the sanctity of the family. The heroes of road-movies, unacknowledged gays or not, are usually determinedly free-wheeling bachelors; Coogan follows in the footsteps of previous Hollywood sheriffs in having no family connections to slow down his manic pursuit; Bickle's solitariness may well be the result of his being trapped catatonically between the twin stereotypes of goddess and whore; and the tangled skein of divorces, affairs, and cohabitations over which Woody Allen nervously presides is a more complex rejection of (or inability to live within) the nuclear family. The New Left and feminist stress on the political aspects of personal life, the post-1968 experimentation with life-styles not based on the nuclear family (such as communes or *ménages-à-trois*), and the work done by political philosophers and psychoanalysts on the oppressive structures of patriarchal society led to an appreciation of how important the nuclear family could be as an organization for exploitation and tyranny.

The film which perhaps best demonstrates the links between the organization of family life and that of social and economic oppression is *The Godfather*. Here the paradox that many violent and ruthless gangsters are also kindly and considerate family-men turns out not to be a paradox at all. Coppola's decision to make a film about the Mafia (though they are never actually referred to by name within the film) was commercially motivated, but it also enabled him to unfold two narratives simultaneously – one of organized crime on a scale that makes Fuller's *Underworld USA* appear positively amateurish, the other of Catholic family sacrament and ritual – which the film's ending confirms as parallel. We have seen a tapestry of businesslike violence framed by two sacraments, a wedding at the beginning, a baptism at the end; we have seen Marlon Brando/Vito Corleone fall victim to violent revenge, not in the middle of some criminal undertaking but in his role as family-man going across to a stall to buy fruit; we have then seen him drop dead while playing with his grandson in the garden; and we have seen Al Pacino/Michael Corleone, the best-educated and 'straightest' member of the dynasty, succeed to his father and ruthlessly avenge his death in the name of family solidarity as much as of underworld efficiency. The cutting between the baptism of Michael's godson and the hecatomb of his rivals acts to establish both as in a very real sense family sacraments.

Diane Keaton/Kay Corleone is unhappy about her husband's suspected involvement in the underworld he had always sworn to abjure. Coppola's original ending was to have shown her lighting candles in a church, but that included at the behest of the film's co-producer, Robert Evans, is immeasurably more effective. Kay receives an assurance that Michael is in no way involved with crime, and the film shuts the door in her – and the audience's – face. Michael and the film ironically combine to deny what the past three hours have been devoted to proving to us, that the organization of the family and that of crime are analogous; and this is why the film's two lines remain, precisely, parallel. They never, within the film, meet. The closing of the door is so blatant a denial of what has clearly been demonstrated that it works against itself. It would be dangerous to make over-generous claims for *The Godfather*'s subversive qualities – it is, in many ways, complicit with much of what it purports to condemn – but looked at in the context of a widespread dissatisfaction with the nuclear family it is very much of its time.

Important though the Vietnam war was for the radical and protest movements of the 1960s, it did not filter through into mainstream American cinema until a good deal later. There was, of course, John Wayne's notorious *The Green Berets*, a direct transposition of the cowboy-and-Indian simplicities of the Western to the Far East, complete with reflective sunset ending. The Hollywood 'establishment' embarrassedly attempted to wash its hands of the film; liberal and radical critics were torn between denunciation of its iniquity and mockery of its crassness. Neither reaction did anything to change the fact that it made an immense amount of money, particularly in small-town cinemas whose audiences reportedly cheered it to the echo. The film now appears as a gloss on the genre which inspired it, flushing into embarrassing prominence a bullying imperialism and a dismissive cultural arrogance (both masked as patriotism) that the time-honoured iconography of the Western has tended to blur. The film's enthusiastic reception may well have flowed from its very simplicity, and Wayne's political astuteness in that case was greater than generally realized; what could have been shrewder from a box-office point of view than to clothe the tortured complexities of a defunct imperialism in the reassuringly self-confident garb of a 'Duke' Western?

The Green Berets, however, was an exception. Major studios were obviously extremely reluctant to tackle the war on film until its result became known. Wayne's proselytizing conviction has meant that his film is now viewable primarily as a joke that left-liberal audiences can share at the expense of its maker. Few studios, producers, or directors were willing to lend themselves to such Messianic agitprop; criticism of and opposition to the war tended to take different cultural forms, notably via rock music; and such overtly anti-Vietnam films as were made were almost by definition situated outside the main commercial circuits. Once the process of 'Vietnamization' had run its fruitless course, and defeat finally had to be admitted in 1975, something like an embarrassed silence characterized the immediate response. The traumas – of defeat by a humiliatingly smaller opponent, of agonizingly pointless deaths, of the political and ideological ramifications – were too many and too complex to be readily digested (especially by a nation still reeling from the aftermath of Watergate).

1978 and 1979 marked the turning-point. Three major box-office films of those years addressed themselves to the problems of Vietnam – not so much the overtly political implications of the war as its impact upon the community life of America. To postulate anything so unified as a cross-studio agreement on how finally to present the Vietnam conflict on film would be to indulge in conspiracy theory on the truly grand scale; yet, viewing *Coming Home*, *The Deer Hunter*, and *Apocalypse Now*, it is difficult not to see running through them a common thrust towards depoliticizing the war, not merely through (at best) ambivalence on the justification for the American intervention but through a recasting and digestion of it into established movie genres which help to neutralize much of its charge. And, of course, such a formula turned out to be superb box-office; the one issue no American consciousness of the previous ten years could have escaped combined with the presence of well-known stars and directors and the use of tried and trusted genre forms to prove immensely lucrative.

Coming Home, of the three films the one most unequivocally critical of the war, recasts it into a 'woman's-picture' format. Our brief look at Sirk has shown how much subversive charge this genre can carry, but in *Coming Home* this tends to work against rather than with the already highly charged subject-matter. Jane Fonda/Sally Hyde is married to a self-confident US army captain; it is during his absence on combat duty that she experiences the first orgasm of her life, with a paraplegic veteran turned ardent anti-war campaigner. The symbolism (of different forms of paralysis, of war and love as mutually exclusive uses of energy, of the global failure of American masculinity) comes across as too stereotyped and small-scale for the breadth of the issues invoked. What would have worked perfectly well in the low-budget small-town America of the 1950s cannot stand up at a later period when larger questions – and sums of money – are involved. The film appears indeed as a none too coherent patchwork quilt of pieces borrowed from three generations: the woman's picture from the fifties, the time of the narrative (underscored, often tediously, by one of the archetypal sixties-nostalgia soundtracks), and the embarrassed liberal guilt with which much of America faced up to the aftermath of Vietnam in the seventies.

Not that there is anything particularly surprising in this; the manner in which films fail in one way or another to cohere is, we have seen, often as interesting as that in which they succeed, and it would hardly be realistic to expect Hollywood, after its period of hibernation, immediately to have produced a fully cogent text on the implications of the American defeat. But, as Richard Combs says, the film 'gives the feeling that it has set itself the epic task of putting together all the pieces of post-war America',[34] and this its genre format largely precludes it from doing. The woman's picture can expose major conflicts and contradictions where none is popularly supposed to exist, but its suitability for dealing with struggles of near-global proportions is questionable. It is disquieting to consider how widely *Coming Home* was accepted as a kind of Hollywood act of contrition for the war, when its major thrust is towards suggesting that the Political (capital letter) is but the personal (small letter). In Hollywood terms, Ashby's film is the anti-Sirk.

Coming Home, the woman's picture, corresponds, however adventitiously, to the 'man's picture', *The Deer Hunter*. The main cinematic references here are to John Ford Westerns (as in the final singing of 'God Bless America'); the wider cultural context thereby invoked is – at first sight – an oddly anachronistic one. It is the world inhabited by Hemingway's warriors and hunters, that of André Malraux's 'virile fraternity', a universe of male communion whose sacraments are guns, meat (hunted rather than bought) and alcohol – all commonplace in *The Deer Hunter*. What is the relevance of such a treatment of the Vietnam conflict?

In the first instance, it has to be recognized that this world is anachronistic only from a detached and minority urban perspective. The various hunting rituals in the film do still form an important part of much American small-town life, the more so perhaps as increasing mechanization at work increases their recreational value by contrast. This said, however, one is on shaky ground in attempting to justify their place in the film on 'realist' criteria; what, by those same criteria, is the justification for the notorious Vietcong Russian roulette sequence, for which no historical evidence has been forthcoming? It seems to me that a further-reaching justification lies in the paradox explored by John Pym in his article, 'A Bullet in the Head: Vietnam Remembered'.[35] This is that Vietnam was, on the one hand, a faraway country of which most Americans knew little, even when the names of its main cities and geographical features were household words, and, on the other hand, directly and unceasingly available, night after night, through the high-technology filming of a high-technology war. The noisy and inescapable reality of the conflicts and the unknownness of the culture combined to make possible a filmic world in which such distortions as the Russian roulette sequences can call upon the stable and ordered rituals of an older America to correct and contrast themselves. The deer-shooting and the wedding are the other pole, not just to the turmoil of the steel-works, but to the turmoil of Vietnam; and this is true, not just at the fairly self-evident level of ethics and loyalty, but also at that of technology. The notion that technological sophistication necessarily results in greater accuracy and pertinence of communication has probably never been more fully rebutted than by the newsreel presentation of Vietnam. *The Deer Hunter* goes to show that it can also result in a 'de-realization' of just those details it most vividly foregrounds.

The three friends, Michael, Nick, and Steven, are prisoners of the crisis of imperialism in more ways than one. Not only does it never occur to them (as it must never have occurred to hordes of volunteers, in the war's early days at least) to question the rights and wrongs of an American intervention in which they are proud and excited to participate; between the set-pieces of rituals which seem immutable because they do not depend on high technology, and the sound and fury of a conflict amplified the better to deafen them to its deeper resonances, they are trapped.

This is most obvious for Nick, whose brutalization by the Vietcong leads him to become a heroin addict and professional Russian roulette player – in some sense the logical consummation of the war as grisly media spectacle, as well as a hideous parody of the shooting rituals of his Pennsylvania days. It may seem less so for Michael, whose inability to shoot a buck on his return from Vietnam appears to bespeak a chastened gentleness further exemplified by his final ability to make love to Linda. But this episode too is ambiguous – has he become vegetarian or cannibal? After all, when he returns to Vietnam and finds Nick, he is able to mouth the words 'I love you' – a final admission of the combined intensity and hopelessness of 'virile fraternity'? – yet quite unable to prevent Nick from finally blowing his own brains out. It is almost as though Nick's only possibility of reintegration into the small-town society by this point is as a corpse, for whom 'God Bless America' can safely be intoned to bring the ritual round full circle.

The Deer Hunter, in other words, works on confusion rather than contradiction. Its epic façade glorifies all kinds of admissions of defeat and hopelessness while itself symbolizing what it constructs as the only possible alternative – ever larger and more ambitious media coverage and representation. The film's most culpable area of confusion, however, must surely be its blatant eviction of the historical. The three young men set off as exuberant volunteers sometime in the sixties; Michael's final return is just before the fall of Saigon in 1975. The intervening period had seen a vast upsurge of criticism of the war (based on disgust, weariness, bereavement, or all three), of which no sector of American society could have remained unaware. Yet the film ignores it completely. Only a major historical trauma could have justified the making of *The Deer Hunter*; but the trauma is retained, recuperated in the form of an adventure-story whose protagonist returns a sadder-but-wiser man, and the history either treated as though it had never existed ('God Bless America') or reduced to an uneasy hybrid of anecdotal distortion and fuzzy allegory. Are the Russian roulette scenes 'merely' directorial licence or an oblique comment on the dangers of any kind of flirtation with 'Communism'? Either way, they are as obnoxious as the rest of the film.

Vietnam as spectacle looms even larger in *Apocalypse Now*, where the 'Ride of the Valkyries' sequence places us in an American helicopter and treats us to the euphoric excitement of a rocket attack. The film is strewn with archetypes, literary (allusions to T. S. Eliot and Conrad), anthropological (Sir James Frazer – interestingly, an early anthropologist whose work is now unfashionable for its tendency to itemize at the expense of analysis), and psychoanalytical (Martin Sheen/Willard's journey upriver towards Marlon Brando/Kurtz is a kind of Oedipal odyssey in reverse), which work to broaden its reference but at the same time to obfuscate it. The sense is of a trauma so overwhelming as to require a quasi-divine frame of reference, and at the same time majestic in its very dimensions. Even more than in *The Deer Hunter*, the influence of newsreel filming of Vietnam – the blowing-up of incidents and locales so as to 'de-realize' them – is apparent in the river and bombing sequences.

Kurtz when he finally appears seems to have more to do with the monstrous American dream of *Citizen Kane* than with any historical conflict. The episode where he shambles into Willard's bivouac holding a ball-shaped object which turns out, when he drops it into Willard's lap, to be the head of his colleague Chef, in some ways echoes the 'Hall-of-mirrors' Rosebud sequence near the end of *Kane* – visually, to be sure, but also in its evocation of the near-superhuman individual, ground and goal of the 'American dream', driven to madness and senility. There are also echoes of *King Kong* in the monstrous way in which Kurtz lords it over his jungle fastness – but this monster will never return to the big city in chains . . . The point here is not merely that Coppola has brought to bear upon the Vietnam conflict parts of the whole apparatus of Western cultural supremacy that its result called into question, but that this whole tissue of reference and allegory again works to evict history from the film. The archetypal and anthropological works over which Kurtz pores all situate themselves in some way before (or outside) history; Kong is a living denial of the historical, which is one reason why he has to be contained and destroyed rather than being left in peace; and Kane's endeavours to make and, when that fails, to rewrite history are what give him such extraordinary stature.

The one scene where the historical reality of the war intrudes most inescapably is paradoxically that involving the go-go girls, helicopter-lifted into the camp to put on a show whose tawdriness is indisputable. Here we are at the level of the sordidly exploitative, so that even a 'gung-ho' audience of *Green Berets* fanatics, liable to be moved to cheer by the Wagnerian grandeur of the bombing sequences, would be likely to experience this scene as a distasteful anticlimax. Up to this point how we read the film – as superhumanly exciting or superhumanly evil – will in all probability have depended on our previous ideological 'programming' (and the distribution of *Apocalypse Now* will have ensured that it will have played to audiences of all possible shades of opinion). Now, the tattiness of surrogate sexuality upriver briefly intrudes – a reminder of the smaller-scale

American life most of the soldiers will have left behind. With the bombardment and the girls' evacuation, the superhumanly superhistorical returns, and our final foothold in the sociohistorical reality of America's invasion, cultural as well as military, disappears.

What is important across these three Vietnam films, returning to our concept of the 'organization' as articulated at the beginning of the chapter and developed thereafter, is their rewriting of the war in such a way as to exclude its explicitly political elements, and the use each of them makes of earlier Hollywood forms together with the way in which Vietnam was televised to do this. *Coming Home*'s inversion of the operation of the women's picture; *The Deer Hunter*'s alternation of timeless Fordian America with big-scale high-technology Vietnam footage; *Apocalypse Now*'s taking the use of such footage in every sense over the top, and anchoring it in the mythical Hollywood world of, among others, Kong and Kane – all show how the organization of genre, of stars and technology within and across genre, can be used to skirt round political implications still difficult to face up to, and to ensure lucrative audience response. The awesome is an easier option for big-budget Hollywood than the embarrassing.

The theme of the individual odyssey also looms large in all three films. Sally Hyde's journey towards sexual fulfilment, her husband's away from confidence in his own warrior masculinity, Nick's gamble with madness and Michael's return from technological violence to timeless ritual, Kurtz's rebellion (in his own way!) against the course the war is taking and Willard's journey to find him – all in different ways pit individuals against various forms of organization, familial, military, technological, even the different forms of psychic organization collectively lumped together under the heading of sanity. This is, no doubt, why they have been the focus of much debate about their ideological implications. But beneath their different façades of subversiveness it is the organization of Hollywood – at work in the launching and funding of the films, in their narrative structure (what is Nick's suicide but the *nec plus ultra* of the free-market gamble, what Kurtz's insane realm but that of entrepreneurialism run literally riot?), and in the stars and genres they deploy – that is constantly at work, constructing a nation which has method in its madness as well as madness in its method. Which we choose to privilege is an even more difficult question to answer than the others this section has raised, if only because historical distance is lacking. How – or indeed whether – *Apocalypse Now* will be readable in twenty years' time is a question it would require the hubris of a Kurtz to answer.

Part Two

History in the Writing – French Society through its Cinema

French nationality is quite a windfall: it confers upon its possessors a double membership, of the human race as a whole and of one among many nations. This miraculously self-evident fact is rooted, not in everyday logic, but in a particular political situation. Since it was the French nation that historically gave birth, at the end of the eighteenth century, to the very idea of a nation . . . and since 'every man has two countries, his own and France', the religion of the nation is equivalent to a religion of mankind.[1]

Régis Debray's ironic exposure of the political and historical factors that underlie implicit French cultural claims to universality will have a particular value in our study of French society through its cinema. Capital letters, Debray reminds us in the chapter from which the above quotation is taken, denote universals or abstractions, following which it would be possible to say that if the United States (represented by Hollywood) is cinema, France (represented by Paris) is Cinema. The capital stands at once for universality (nowhere else is it possible to see so many films from such a wide historical and geographical compass as in the Latin Quarter of Paris) and for abstraction (theoretical discussion of and writing about film has a longer past and a more vigorous present in France than in any other Western country). The cinema may have been invented in France, but it was not long before the French industry fell behind the American in productivity and profitability, never to recover its position. Nevertheless, cinema as object of discussion, dissemination, and study is as surely localized in Paris as cinema as object of glamorization and investment is in Hollywood. Not the least important aspect of the reproduction of French society in its films is the awareness that film-consciousness is one of its more distinctive strands, among the Parisian intelligentsia at any rate – a point that will assume its full importance when we come to look at the 'New Wave' directors of the fifties and sixties.

But the movement between historically specific and mythically universal to which Debray alludes needs also to be looked at in French society's representations of itself in areas other than the cinematic, if its importance in the cinema is to be sufficiently grasped. The French Revolution of 1789 was the result of a specific set of socio-historic circumstances, notably a corrupt and inefficient monarchy, a feudal order that concentrated a third of the country's land in the hands of the clergy and the nobility, the consequent impoverishment of the peasantry, and the rise of bourgeois productive forces that could find no outlet under feudalism. It led to the emergence of the first bourgeois nation-state, with a declaration of citizens' rights, disestablishment of the Church, elected assemblies, and administrative reorganization (all gains soon to be imperilled or eradicated by aggression from without and repression within). This became a model on which other movements of national reorganization were subsequently to pattern themselves, thus also a historical (and at times transhistorical) archetype. Marx recognized its archetypal status in the very title of his *Eighteenth Brumaire of Louis Bonaparte*, with its parallel between Napoleon's 1799 *coup d'état* which completed the French counter-revolution and Louis Bonaparte's analogous annexation of power after the thwarted socialist uprising of 1848. Already by 1852 the Revolution of 1789 and its aftermath had become the model by comparison with which crises in and of the bourgeois state were judged.

Debray's thesis that the Revolution led to the implicit equation of France with the archetype of the modern nation-state finds confirmation, perhaps not surprisingly, in the discourse of General de Gaulle (himself often regarded as the 'second Napoleon' Louis Napoleon farcically failed to be). Jean-Marie Cotteret and René Moreau, in their statistical analysis *Recherches sur le Vocabulaire du Général de Gaulle*, demonstrate that the ten commonest words in de Gaulle's radio and television broadcasts between 1958 and 1965, in descending order of frequency, were (in their English translation): France, country, Republic, State, world, people, nation, progress, peace, future. About the first on the list there need be little surprise; but what is interesting thereafter is the juxtaposition of global abstractions (country, world, people) and references to specifically French institutions (the capital letters for 'Republic' and 'State' make it clear that it is the Fifth Republic and the French State that are referred to), undershored in the final three words on the list by forward-pointing abstractions in such a way as to suggest that these desirable qualities are quite literally dependent on the nationhood of France.

A specific example will help to illustrate the point. In 1962, shortly after the signing of the Évian agreement which gave Algeria its independence, a group of disgruntled French-Algeria supporters attempted to assassinate de Gaulle at Le Petit-Clamart, just outside Paris. His response – a brilliant political coup in the turning of danger to his own advantage – was to put before the nation a referendum on the election of the President by universal suffrage, instead of as previously by an electoral college. One of his television speeches (quoted in Harris and de Sédouy's film *Français, si vous saviez . . .*) referred to the measure as follows:

> Frenchwomen, Frenchmen, the proposal I put before you is that the President of the Republic, your president, shall be elected by you yourselves. Nothing is more republican. Nothing is more democratic. I would add that nothing is more French, so clear, simple, and straightforward is it. Once more the French people will make use of the referendum, that sovereign right which at my instigation was granted to them in 1945, which they likewise recovered in 1958, and which since then has enabled the Republic to equip itself with viable institutions and to settle the grave problem of Algeria. Once more, the result will express the nation's decision on an essential subject.[2]

The nub of the linguistic argument here is clearly the audacious equation of 'republican' and 'democratic', with some transcendental quality of 'Frenchness', itself in turn equated with the 'clear', the 'simple', and the 'straightforward'. This is ironic considering how fratricidal and imbroglio-ridden much of French republican history has been; but it enables us to see how de Gaulle was able to coopt, amplify, and extend the archetype of the 'French nation' – first among equals, trail-blazer of modern nationalism, and distillation of common-sense lucidity.

The specific effect of this perception upon films of the Gaullist and post-Gaullist period will be examined in due course. More generally, it helps to account for the rather different emphasis that major historical events tend to receive in French as against American cinema. Such a generalization is perilous, not least because of the difficulty of defining exactly what one means by 'French' and 'American cinema'; but its heuristic value will, I hope, become plain in the following example.

Most of the fifteen films signed by Jean Renoir in the 1930s were contemporary comedies or dramas, dramatizing the social conflicts and tensions of the period without reference to specific historical events (for example, multi-layered class clashes in *La Règle du Jeu*). But there are three other important categories. Firstly, literary adaptations whose contemporary relevance may well be occluded by their literary label. Thus, *Une Partie de Campagne* was made just after the left-wing Popular Front government had instituted paid holidays for workers, which probably helped to account for Renoir's decision to film the Maupassant short story about a bourgeois day-trip into the countryside. Secondly, a specially-commissioned propaganda film, *La Vie est à Nous*, made for the Popular Front before the elections of 1936. And finally, two dramas set in earlier periods – *La Marseillaise*, a 'historical reconstruction' of the French Revolution, and *La Grande Illusion*, set in a prisoner-of-war camp during the First World War. These last two are the most interesting from our point of view here, for both are anchored by historical references in the time about which they are made (the march on Paris in *La Marseillaise*, the battle around Fort Douaumont in *La Grande Illusion*), yet both have at the same time a direct relevance to the time of their making. *La Marseillaise* was an attempt to coopt French nationalism (with its concomitant implications as outlined earlier) into the service of the progressive, anti-Fascist cause: *La Grande Illusion's* prescient analysis of the precariousness of Franco-German relations and the horribly tangible artificiality of European national boundaries speaks, to a present-day audience at least, as eloquently about the late thirties as about the First World War. René Prédal says that *La Grande Illusion* is: '. . . historically far more revelatory as a film about the French state of mind in the years immediately preceding the Second World War than as a document about 1914 – 1918.'[3]

It can equally well be said that *Red River* is nowadays more revelatory as a document of the sexual and territorial apprehensions of post-war America than as a record of what it was like to go on a cattle-drive; but the crucial difference between the Hawks and the Renoir films is that the former takes place, as we have seen, in the quasi-atemporal Utopia characteristic of the Western, whereas both *La Marseillaise* and *La Grande Illusion* have definite geographical and chronological roots in French history. They represent, in other words, a rewriting of that history into the time of their making, much as our analysis here will represent a further rewriting into the present day.

This is not to say that American cinema does not often undertake a comparable rewriting (the retrospective depoliticization of Vietnam analysed at the end of the last chapter is a striking example), but rather that the dialectic between the time about which a film is (ostensibly) made and the time of its making often comes into sharper focus in French cinema because of the greater weight frequently accorded to specific historical events, and – just as importantly – because of the tendency to transmute history into myth whose importance in France is clear. It is no exaggeration to speak of a political nostalgia industry, analogous in function to if less lucrative than its American counterpart, the cultural nostalgia industry. What George Lucas did for the sights and sounds of 1962 popular culture eleven years later in *American Graffiti*, and what Chris Marker did for the political upheavals of 1968 ten years later in *Le Fond de l'Air est Rouge*, have much in common. Both directors attempted to articulate the styles (Lucas) or events (Marker) of a decade before in such a way as to make the maximum impact upon their 1970s audiences, whether by emphasizing difference and thereby recommercializing bygone fashions in music and dress (*American Graffiti*), or by pointing out similarity in the hope of reinvesting the latent radical and revolutionary impulses of 1968 in a society whose basic ills and contradictions had hardly changed since (*Le Fond de l'Air est Rouge*). If the American film's founding myths were pop-cultural

('Rock'n'roll's never been the same since Buddy Holly died'), and those of the French film political, this is because of the privileged place of politics within French society, itself largely the result of the kind of privileged status suggested by Debray.

A kind of 'double vision' will thus often be required in our reading of French society through its cinema, paying attention to the contexts in which historico-mythical events were rewritten and reinvested as well as to the filmic representation of the events themselves. Jean-Patrick Lebel, in *Cinéma et Idéologie*, sees the ideological implications of a film as triply determined: by the elements it selects from 'reality', by their arrangement within the film, and – often forgotten by critics and theorists – by what Lebel refers to as 'contingency'. This relates at once to the physical limitations of filmic signifying material and to the multiple economic constraints upon the film's makers, which at a relatively simple level combine to ensure that, at least since D.W. Griffith, the storming of a walled city cannot be set up and filmed 'literally' (it is thus an important influence upon stylization). 'Economic constraints' needs to be taken as referring, not only to how much money happened to be available for the shooting of the film, but to the need not to antagonize those, backers and audiences alike, upon whom the film might depend to recoup its production costs. This factor in its turn has obvious political impications; the fury with which *La Règle du Jeu* was banned after its first screening, the widespread filmic portrayal of the Maginot Line as unbreachable just before the Second World War, Jean-Luc Godard's conflicts with the French ORTF and other broadcasting authorities over material commissioned from him and then found 'unacceptable', are all rooted in the economic and political conjunctures – the history in its widest sense – of their time, rather than merely in the 'stubbornness' of individual film-makers or the 'stupidity' of audiences. It is these wider factors that this analysis will attempt to elucidate.

The Early Days

The first films ever shown in public, in Paris in 1895, were made by the Lumière brothers (Louis and Auguste) to help publicize the photographic equipment they manufactured in their Lyon factory. It is thus not surprising that their programmes of short films should display, to quote Marc Ferro, 'the creations of the rising bourgeoisie: a train, an exhibition, republican institutions'.[4] Events assume greater importance than individuals; with a few exceptions, such as the malicious urchin in *L'Arroseur Arrosé*, character-study is minimal, and the camera's function appears to be neutral observation rather than the manipulation of narrative and atmosphere. Such a formulation is of course misleading, for so-called 'neutral observation' rests upon selection and arrangement of material just as much as any more obviously interventionist strategy, and the 'realist' qualitites of the Lumière films are far from innocent. It is interesting to view them with their original publicity purpose in mind; the half-proud, half-sheepish attitude of the delegates to the photographic congress whose riverboat excursion is filmed then becomes an oblique homage to the new medium's powers and to the efficacy of the Lumière products, capable of capturing nuances of spontaneous response and of producing for their purchasers and users *photo-*, if not *ciné-, vérité*. The connection between the rise of the bourgeoisie and realism in literature has frequently been made; in the Lumière films we see a similar connection at work in the domain of the image.

Georges Méliès, who began to produce his extravagant but studio-bound fantasies just after the first Lumière screenings, is often contrasted with the Lumière brothers, in terms of the fantastic versus the 'real', the staged versus the observed, or the imaginative versus the documentary. Such statements compound imprecision with inaccuracy. For Méliès was not merely the first science-fiction director, with *Voyage dans la Lune* among others: he also produced a whole string of 'documentary' films which were in fact studio reconstructions, of such events as the coronation of King Edward VII of England, as well as venturing early into the newsreel market (he bid in 1896 to cover the visit to France of Tsar Nicholas II). Even his 'fantasy' films are solidly rooted in the imagery and décor of turn-of-the-century France; the female moon-dwellers in *Voyage dans la Lune* look remarkably like Montmartre chorus-girls, the mad inventors who recur in a number of films (often providing the initial narrative impetus via the crazy craft they design) would not have appeared nearly as fantastic in the technologically expansive nineties as they do now. Méliès himself, after all, was their close cousin . . .

Again, as with the Three Little Pigs and King Kong, we see that cinema as fantasy and cinema as document are not opposite, but different sides of the same coin.

The documentary value of film became evident with the outbreak of the First World War, and the setting-up of the Army Film Services. The events of wartime became the focus for the first large-scale European documentary filming, four major operators (Pathé, Gaumont, Éclair, and Eclipse) covering the same episodes simultaneously, and often incurring the displeasure of the front-line troops. Much of what we now accept as part of the cinematic institution had its genesis in wartime: the 'balanced' variety of coverage by competing operators, the machinery of governmental control (censorship of quite a severe kind was introduced in 1917) and the propaganda use of the medium. Pacifist films were quite common before the outbreak of war, (the Belgian Alfred Machin's *Maudite soit la Guerre!* was premièred only a matter of weeks before the Sarajevo assassination), but they swiftly gave way to patriotic uplift once hostilities had started.

Conditions at the front could not be depicted in too unfavourable a light; Pétain insisted on the excision of a newsreel passage which showed him drinking the wine issued to regular soldiers and grimacing with distaste.[5] The medium's positive propaganda value was also recognized, as when Colonel Marchand in a letter to the magazine *Le Film* in 1914 stressed its utility in the colonial enterprise, referring to it as 'obviously the weapon which will conquer Africa and many other places'. Marchand was not alluding simply to the idealized representations of life under colonialism that film could purvey, but on a more basic level to the cinema as 'magic lantern', as illustration of the white man's shamanic powers. The first cinematic screening in Fez (then under French rule) in the same year followed closely on a guerrilla attack on French soldiers; the film-show, of newsreels accompanied by the Marseillaise on a wind-up gramophone, appears to have had an almost miraculously reconciliatory effect, to judge from a speech delivered by M. Demaria to the banquet of French cinematic unions: '... those very Moroccans who, a few months before, all more or less complicit with an odious massacre, had our soldiers murdered in cold blood, greeted the emblem of France with their applause.' However inaccurate this account may be, what emerges strikingly from it is the early belief in the inherent efficacy of film as an agent of national propaganda, through its apparent combination of 'magic' and 'realism' – a combination which I believe to lie at the root of any initial fascination with cinema. As ninety per cent of films shown in the world in 1914 were French, it is not surprising that the industry's self-confidence at this stage was so high.

After the war was over, and the senselessness of its slaughter had begun to be appreciated, films critical of it reappeared. As a general rule we shall discuss films chronologically according to the time when, rather than about which, they were made, but one interesting exception (details of which again are quoted by René Prédal) merits consideration here. This is Raymond Bernard's adaptation of Roland Dorgelès's novel *Les Croix de Bois*, shot in 1932 in conditions as close as possible to those of 1914–1918. The early and middle thirties, before the menace of Fascism had become sufficiently clear, were of course periods of strong and widespread pacifist feeling, so that the relatively large number of films attacking the First World War as a paradigm of all international conflict is not difficult to understand. Bernard's film is of interest because of its literal historical realism; seventy-five per cent of the film consists of battle-scenes, which were shot on actual First World War battlefields (often using surviving trenches) and played entirely by people who had taken part in the war. The film was given a special screening to old members of Dorgelès's regiment, whose applause was vociferous. Such retrospective documentaries (cf. later, René Allio's *Moi, Pierre Rivière*) were to be quite an important French sub-genre.

After the war, landscape- and location-filming became a staple of the silent era, persisting well beyond the advent of sound. French social, cultural, and political life is overwhelmingly concentrated in and upon Paris, and the effect of this upon France's cinematic representations of itself will be an important theme of our study. In the inter-war period, the only provinces that according to Prédal had 'a relatively satisfactory cinematic portrayal'[6] were Brittany and Provence – a choice that was not adventitious. Both are popular holiday areas for metropolitans; both abound in 'local colour' (seascapes, villages, customs and traditions) of a kind that has always appealed to tourists, especially those with cameras; both have well-defined stereotypical identities, human and geographical. The quarrelsome, chauvinistic, poetic, and inebriated Breton is seen as the 'natural' product of his remote rocky coastscape just as the jovial, extrovert, pastis-swilling, joke-telling Provençal is seen as that of his sun-soaked, palm-groved, *pétanque*-playing habitat. The grotesquerie of such formulations should not obscure their persistence, whose effects are frequently pernicious; the economic deprivation of Brittany could well have been facilitated by its colourful image, much as the alleged eccentricities of the inhabitants of Britain's 'Celtic fringe' have helped to reinforce patronizing and exploitative attitudes towards them.

The Marseille-based films scripted, produced and/or directed by Marcel Pagnol anchor their charm and warmth in precisely such a stereotype of the 'Provençal character' – mellow and jovial through all vicissitudes and twenty-three years of filming. Pagnol financed Jean Renoir's *Toni*, also set in Provence but a world away from the benignity of Pagnol's studio sets. *Toni*, made in 1935, is a drama about migrant labourers working in a quarry near Marseille, shot on location with a largely non-professional cast. The hero, romantically entranced with Josefa, Albert, her coarse caveman-like husband, and Gaby, the cousin who has been her lover for two years, form a doomed romantic quartet whose boundaries are the river, the railroad beside which Toni is shot down, and the quarry – the worlds of 'natural' environment, of travel, and of work, the three worlds between which the characters move. Work oppresses them literally as well as metaphorically (the quarry-walls engender a powerful sense of claustrophobia); travel figures in their lives only as a prelude to work (the film begins and ends with migrant workers arriving by train), or as an escape from justice (Gaby, arguably an accessory after the fact of Albert's murder, commandeers Toni's motorcycle to flee to Marseille); the river, the aquatic element classically associated with musing and reverie, intersects with the railroad, for as Leo Braudy points out: 'The railroad in *Toni* crosses the river, suggesting the rigid fatalities to which Toni has committed himself because he can envision nothing beyond it.'[7] What is important here for our purposes is not so much the parade of archetypes as the absorption of avenues of escape, by road, rail, or water, into the claustrophobic world of the workers. In this way the location-filming becomes more than a financial expedient; it is indissociable from the film's imagery and poetics, themselves indissociable from the clearly delineated world of the migrant workers.

Provincial location-shooting was comparatively rare in sound features even as late as 1935. Studio filming was far commoner; there is a tendency to use the word 'reconstruction', but this is subject to two criticisms. First, it leaves out of account certain Surrealist-influenced films (such as *L'Age d'Or* and *Zéro de Conduite*), which certainly do not aim at anything like a 'realistic reconstruction' of the décor and society of 1930s France, yet tell us a great deal about the period. Secondly, it implies that the studio sets were pallid imitations of their 'real-life' originals, intrinsically inferior to shooting on location – a belief fostered by much realist writing on the cinema, notably that of André Bazin. It is interesting in this connection to look at Annie Goldmann's comment on French cinema in the thirties: 'It would be possible to analyse the French cinema of 1930-1939 in terms of its three main currents: the anarchistic one of Jean Vigo and René Clair, the romantic one of Marcel Carné, and that of Jean Renoir, linked to the ideology of the Popular Front.'[8] The division is certainly a tendentious one, but it is worthwhile noting that of the three currents Goldmann identifies, the first two are associated with studio- and the third with location-filming, which at least suggests that the two modes of working produced distinctively different results. Vigo and Clair, lumped together under the heading of 'anarchistic', both have some claim to the label, but it hardly does justice to their very different views of French society, nor to the presence of 'romantic' elements (perhaps a necessary corollary of anarchism) in their work.

What can be seen in both directors' films is a romantic aversion to bourgeois conventionality, and especially to established bourgeois ways of earning a living. The central characters of Clair's *Le Million* are a sculptor and a painter, both pursued by creditors – a scenario familiar to readers of Balzac or audiences of *La Bohème*, and one timelessly inscribed in literary and artistic representations of Paris up until the crisis of the thirties and the Second World War. Thereafter, young men of genius (Bohemian geniuses were always male) went down to the streets to drink beer and demonstrate or into the cafés to drink coffee and discuss philosophy, instead of remaining in garrets to starve and create.

Le Million is interesting now as one of the last gasps of

Parisian Bohemia, which is signalled by the elaborately choreographed scrimmage around the jacket containing a winning lottery-ticket and the happy ending (the artists win the million francs to which they are entitled). Old-school Bohemia never took itself with such levity, nor allowed itself the comfort of such happy resolutions. Along with Bohemia there coexists a kind of urban Arcadia – pastoral in a tenement lodging-house – dependent for the sense of joyous community it constructs on the use of a studio-set. Clair's Paris, here as in such earlier works as the musical *Sous les Toits de Paris*, is a romantic distillation which could not have survived topographical precision. It is 'everybody's Paris' precisely because nobody's.

Unlike *Le Million*, *A Nous la Liberté* does show people doing an 'honest job', but in an obviously critical manner. The factory workers, through the recurrent bars motif and the grim standardization of their production-line, are visually equated with the prisoners at the beginning of the film. Their boss, named Louis, a good capitalist in that he clearly considers himself philanthropic because of all the work he is creating for others, is an escaped convict, detected only when he has reached the pinnacle of achievement. It is not surprising that Hungary (then under a right-wing Regency) and Portugal both banned the film, and that Mussolini permitted it in Italy only after the title had been changed to *A Moi la Liberté*.

To read the film just as an anti-industrialist tirade, however, is to weaken its force (the ending, in which the boss makes over his factory to the workers in it and sets off down the open road with his old cell-mate, Émile, is politically feeble), and to disregard its subtler exploration of the mechanisms by which pleasure is produced. The factory workers assemble gramophones, at that date the latest popular form of home entertainment. When 'released' from their servitude by the 'governor' of their 'prison', they are shown indulging in various open-air pursuits – fishing, dancing, cruising, lounging – in lines and ranks as ossified as any they formed inside the factory. Lines and bars recur constantly throughout the film, culminating in the departure of Louis and Émile at the end along a straight road towards the horizon. Their song tells us:

> Everywhere, or so I am told,
> Everywhere you can laugh and sing,

Female careerism in the sale of the body-as-spectacle: Lucille Ball in <u>Dance</u>, <u>Girl</u>, <u>Dance</u> (courtesy of the BBC)

America

Snared!... Cary Grant trapped by Katharine Hepburn in Bringing Up Baby
(courtesy of the BBC)

'VOTE FOR ME!' Orson Welles in <u>Citizen Kane</u>
(courtesy of the BBC)

Gary Cooper doing what a man's gotta do in High Noon
(courtesy of the BBC)

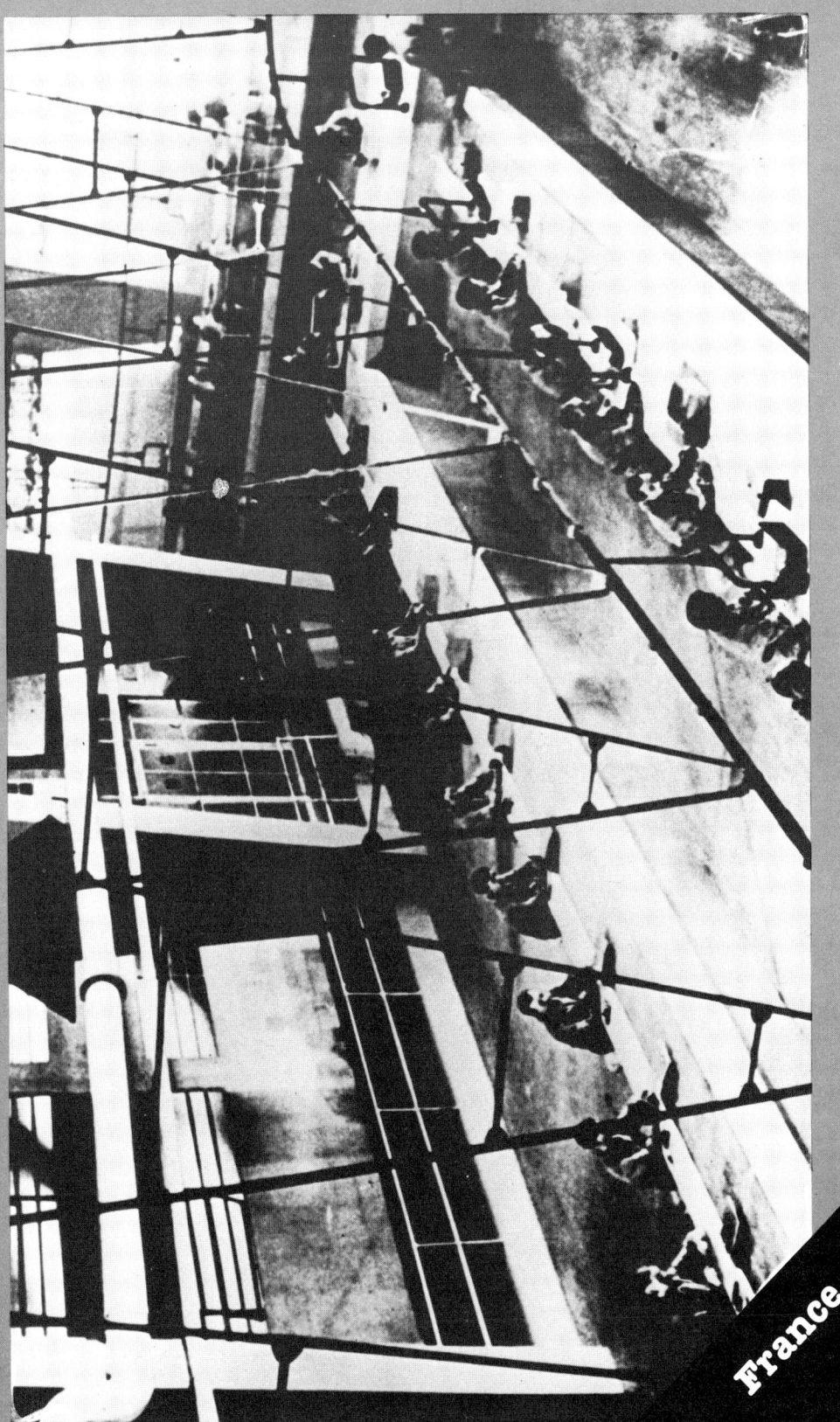

The automated production of entertainment, in *A Nous La Liberté* (courtesy of Marceau-Cocinor)

France

Death on the Riviera: Jean-Paul Belmondo and Anna Karina in <u>Pierrot le Fou</u> (courtesy of Bela Productions)

a) Death for the voyeur, Carl Boehm in Peeping Tom (courtesy of EMI Films)

b) Class revenge as Dennis Price catches Alec Guinness in the mantrap, from Kind Hearts and Coronets (courtesy of EMI Films)

Great Britain

Japan

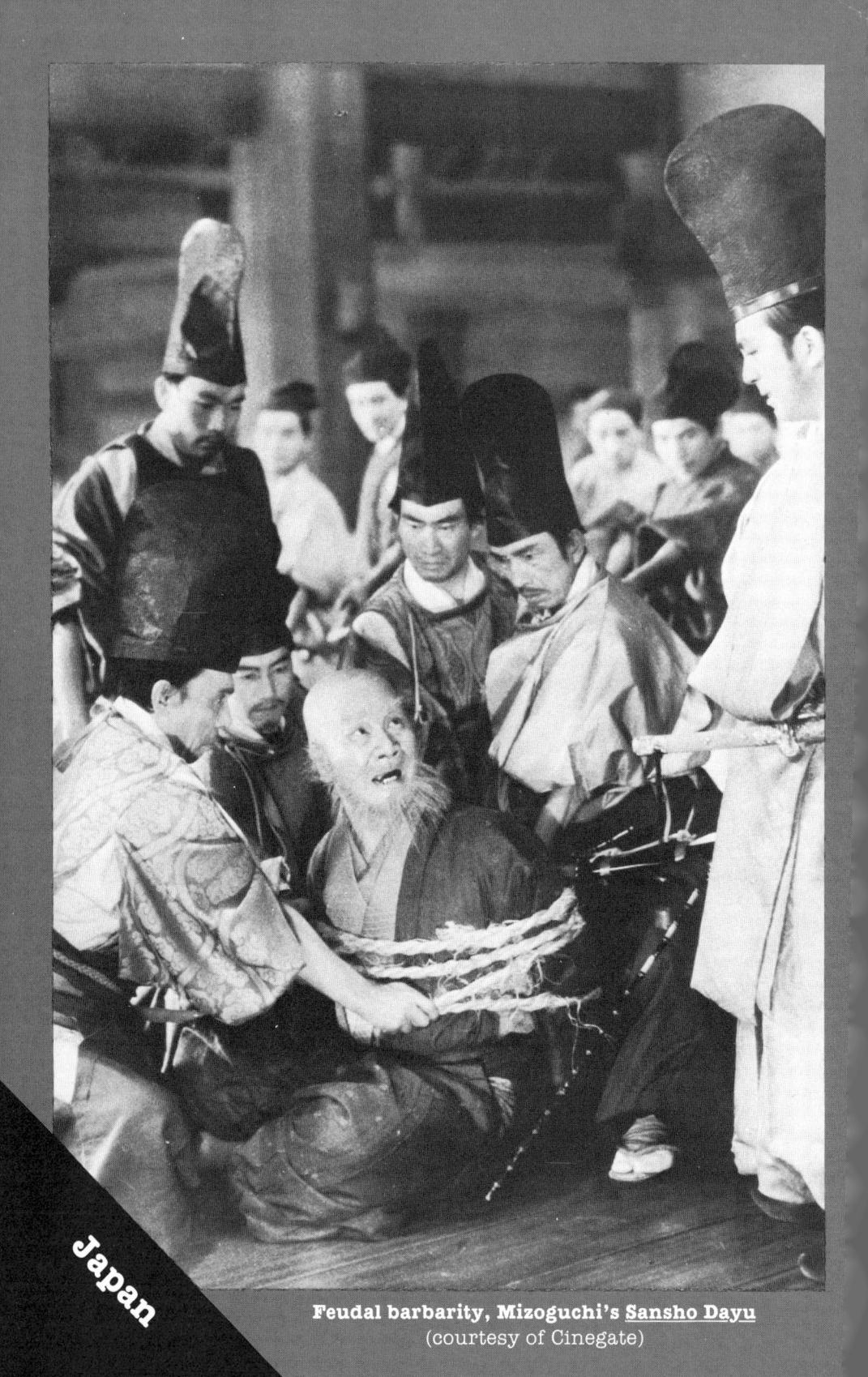

Feudal barbarity, Mizoguchi's Sansho Dayu
(courtesy of Cinegate)

Everywhere you can love and drink,
Freedom for us![9]

'Is that all?' a cynic might be tempted to remark. The 'alternative' vision implicit in Clair's narrative resolution is one of simple pleasures liable to be dulled for the bosses (and it would seem dulled in advance for the workers) by repetition. Freudians will remember that the apex of the pleasure-repetition triangle is death. Without trying to present *A Nous la Liberté* as a morbid or fatalistic work (which would be absurd), I nevertheless consider it important to point out how its most subversive thrust, at a level far deeper than the rather fey critique of big factories and small people, is at the limitation and standardization of ideas of pleasure under totalitarianism.

The film's iconography is clearly pre-Fascist (Louis the boss addresses his workers from a podium in obvious dictatorial style), which is presumably why it appealed to Mussolini despite its anti-authoritarian thrust. Vigo's anarchism diminishes rather than emphasizes the iconography of authority; in *Zéro de Conduite*, the school headmaster is a midget and the notables on the speech-day platform at the end are inflatable (and hence presumably deflatable) dummies. Catholic intervention (the equivalent of the Legion of Decency in Hollywood) succeeded in getting the film banned just after its first screening, its 'anti-French spirit' (like McCarthy's 'anti-Americanism') and Vigo's alleged Communist sympathies supposedly going to influence this decision. Fortunately the censorial process was not so well-orchestrated, nor so ideologically consistent, in the France of 1933 as in McCarthy's America, and film-societies (which had originated in France and were still quite a potent influence there) ensured the film limited distribution.

It may well have been precisely the resolute 'irrealism' of Vigo's film that led to its banning. To show authority-figures, as in Clair's *A Nous la Liberté*, who are not only dangerous (Louis alienates many of his colleagues through his incipient megalomania) but illegitimate (he has arrived at the top under false pretences), was acceptable; to present them in Vigo's way – as literal grotesques whose power and legitimacy reside not in any properties of their own, but in the fear they inspire in others – was beyond the pale. This was particularly likely to be true in the volatile political conjuncture of 1933 France. There had been a Fascist riot against *L'Age d'Or* a few years before, as a result of which the film had been banned; the social-democratic régime was to be seriously discredited by the Stavisky affair, in which a consummate swindler achieved a position of high public financial responsibility, and involved members of the government in some of his deals; this episode helped to spark off an attempted (and very nearly successful) military-Fascist coup in February 1934; and the Communist Party was meanwhile a powerful force on the left, among manual workers and intellectuals alike. Vigo was associated with the Surrealists, whose leading literary spokesman, Louis Aragon, had become an enthusiastic Communist after visiting the Soviet Union in 1930; and his father, the anarchist journalist and militant Miguel Almereyda, had been murdered in Fresnes prison in 1917, in circumstances never satisfactorily explained. It is thus hardly surprising that his work should have been more suspect than that of the more respectable Clair (first French film-maker ever to have been canonized by membership of the Académie Française), on personal grounds alone.

If one looks at *Zéro de Conduite* without reference to Vigo's personal circumstances and reputation, it is still comprehensible that it should have been banned on first release (and that it should still have something of the status of a *film maudit*, usually screened to film-societies or late-night art-house audiences in conditions of subversive anticipation). The centralization and bureaucratic pomposity of the French educational system are mercilessly lampooned, along with the corruption of the masters (one of whom interferes with boys in class) and the abominable food. Nothing was better calculated to elicit howls of chauvinistic rage than a combined assault on the educational system and the institutional catering of the supposed intellectual and gastronomic centre of the civilized world (Debray's comments on French claims to universality are even truer in the field of cookery than elsewhere). It was thirty-five years before the ossification of the educational system was even partially remedied, as a result of the very events of May 1968 which were to influence Lindsay Anderson's *If* . . . – itself an explicit homage to Vigo's film. *Zéro de Conduite* shows clearly how the subversive charge of a film can be specific to the time of its making, yet remain active (for historically determined reasons rather than because of some mysterious quality of 'eternal youth') for a very long time afterwards.

The anarchistic distaste for acceptable bourgeois ways of making a living, already noted in Clair, emerges in *Zéro de Conduite* not only in the corrosive portrayals of the academic hierarchy, but also in the Chaplinesque insolence of the junior master, Huguet, who mocks his colleagues and sides with the boys in their revolt (he clearly has a brilliant academic future behind him and is not in the least concerned). *L'Atalante* takes this a stage further by being set on a barge (an obvious metaphor for the 'free-floating' existence), and constructs an interesting double view of Paris – for the old bargee Père Jules a place of familiar debauchery from which he returns drunk to the boat, for the dissatisfied young bride Juliette a threatening jungle in which she gets lost and has her handbag stolen. Conventionally 'safe' modes of travel are fraught with danger (the bag-snatch occurs at a railway station); the barge and its bizarre population come to represent a haven of security. Something of the urban paranoia of the contemporary *King Kong* imbues *L'Atalante*, darkening its Bohemian anarchism.

The 'anarchist' current in thirties film-making thus identified and highlighted many of the main factors behind the social and political unrest of the period: the precariousness and contingency of authority (moral, economic, or political), the brutalizing effect of urban industrial life (paid holidays were not introduced until 1936), and the sense of volatile discontent that caused the country briefly to waver between the Communist and the Fascist 'solution'. The short life and tortured death of the Popular Front, which came to power in June 1936 and fell in April 1938 after long dissension between Communists and Socialists over whether or not to help the beleaguered Republicans in Spain, and the return to moderate compromise under President Daladier (co-signatory of the Munich Agreement), were followed by the outbreak of war (in September 1939). *Zeitgeist*-hunters, those who delight in tracking down the quintessence of an epoch in its artistic manifestations, have had a field-day with the films scripted by Jacques Prévert and directed by Marcel Carné. *Quai des Brumes* and *Le Jour se Lève*, which date from the immediately pre-War period, exude an atmosphere of fatalistic exhaustion which it is retrospectively easy to equate with the political despair and economic prostration of the France of 1938-9. Such an equation certainly has a good deal of truth in it, but to leave it at such a general level does not greatly help us to examine *how* French society is reconstructed in these films and why their success at the time was so great.

Both films are set in (studio-reconstructed) industrial towns: *Quai des Brumes* in the seaport of Le Havre (destroyed by bombing in the war, which gives the film a fascinating but obviously spurious 'documentary' appeal to a modern audience), *Le Jour se Lève* in an unidentified town, probably in the North of France, and certainly the antipodes of Pagnol's Provence. Both star Jean Gabin, at this period in the cinema the archetypal French working-man, whose screen persona came to connote an honest, rugged reliability often doomed by the confusion and duplicity of the world around him. This may well have been linked to the growing political and economic confusion of the thirties, though the consistency of Gabin's persona, and his dominance in the cinema of the period over (for example) the equally proletarian but wilier and more resilient Julien Carette, might be thought to give this reading a fatalistically reactionary cast; why should major French studios so systematically have favoured the representation of the 'ordinary working-man' as too bewildered to have any control over his own fate? Both films, like *A Nous la Liberté* but less prominently, represent the world of work as one of demeaning toil (doubtless the contribution of Prévert rather than Carné). Gabin has deserted from the army (for reasons never explained) in *Quai des Brumes*, so that his death can be read by really determined *Zeitgeist*-hunters as a precognitive warning of the dangers of pacifism; in *Le Jour se Léve* he works as a sand-blaster, in utterly degrading conditions. Physical escape, a staple of the American cinema of the time (for example, the runaways in *You Only Live Once*, the examples of Bogart/Rick and Wayne/Dunson already referred to), is never possible. In *Le Jour se Lève* Gabin defies the police from the lonely siege-room in which he commits suicide at the end, in *Quai des Brumes* he is lying dead on the quayside as the liner on which he was to flee to South America sails. The romantic interludes with Jacqueline Laurent and Michèle Morgan which initially appear as a way out of his hideous reality turn out to be just that, but the way out is through death rather than happy union.

It is possible to understand why these films went completely out of fashion (and, cause or effect, of distribution) after the war, and in some ways rather unnerving to contemplate why there now appear to be signs of a Carné revival. The main form of escape the films offer their audience is through passive acceptance of the mechanisms of entertainment. The cinema and other entertainment-forms were to be widely used under Vichy as propaganda and narcotic devices, and before the fall of France Carné's films had played to packed cinemas many of whose audiences must have come from working-class milieux. When Gabin penetrates too deeply into the mechanisms of entertainment – when he becomes involved with the toyshop owner in *Quai des Brumes* or the performing-dog trainer in *Le Jour se Lève* – and learns more about what goes on 'behind the scenes' than he should, the result for him is death. Escape through entertainment, and a warning of the dangers of analysing its workings too closely, is the only way out the films appear to offer.

Francis Courtade sees Gabin's self-condemnation in *Le Jour se Lève* as the result of his 'rejecting the class-solidarity of Clara, the big-hearted daughter of the people, and Françoise, the little working-girl, and of his work-mates who gather beneath his window'[10] – a remark whose rather patronizing tone should not obscure its justice. For Courtade the film is the elegy of the hopes raised by the Popular Front, and it is significant that Carné was among the directors affected by the intensification of censorship when war broke out. *Quai des Brumes* was described in a censorial edict as a 'very beautiful work of art', but simultaneously threatened with banning as 'depressing, morbid, immoral, and pernicious for the young'! What is significant here is not only the ambiguity perceived in the film (and in a different way hinted at by the analysis of it here), but also the particular value ascribed to the cinema in wartime. The edict goes on to speak of young people in wartime as 'more than at any other period left to their own devices, and free to go to the cinema, that easily accessible form of entertainment'.[11] Because of conscription, the use of theatre- and variety-stars to entertain troops, and the risk of black-outs, the cinema was one of the few entertainments readily available in wartime – in London people even crowded into cinemas as refuge from the Blitz – and hence particulary prone to censorship. Renoir's work, for reasons we shall examine, was especially severely hit, but the defeatism of

Carné, along with films mocking or criticizing the military, was also stigmatized.

The German Occupation left film-makers, according to René Prédal, with a threefold choice: submission to the dictates of Vichy, pseudo-historical reconstructions, or retreat into the fantastic. Mercifully few examples of the first strategy have survived, but Prévert/Carné's *Les Enfants du Paradis* and *Les Visiteurs du Soir* belong in the second and third respectively. *Les Visiteurs du Soir* is a medieval fantasy about two lovers turned to stone by the Devil, who at the end whips them in a frenzy but cannot stop their hearts from beating; it has often been read as an allegory of the continued life and resistance of the French nation under Occupation, but this is highly suspect. Nothing else in the film makes even the most guarded allusion to contemporary events, and the attempt to politicize what is basically a metaphysical conceit is uneasily reminiscent of certain readings of Camus's *La Peste*, where the rats that infest Oran and then disappear as mysteriously as they had come are seen as a metaphor for Nazism, and the struggle of the hero Doctor Rieux as 'an idealized reconstruction of the Resistance movement such as Camus and others would have liked it to have been'.[12] In both film and novel, the weakness of the reading is that neither evil nor good have any kind of social or historical specificity; the Devil and the rats come from nowhere and disappear (from the screen or the narrative) to nowhere in particular. Such readings smack more of a collective guilt-complex that may have sought to minimize the trauma of occupation and the unpalatable fact of widespread collaboration by a determined attempt to inject discreet political allegory into metaphysically-based works of art.*

If I have classified *Les Enfants du Paradis* as a 'historical reconstruction', it is not without misgivings, for the film really tells us very little about the theatrical world of 1830s Paris. It operates much more as a rewriting of that period into the time of its making, notably through the character of Lacenaire, who has clear affinities with attitudes prevalent in the aftermath of defeat. His nihilism and tranquil acceptance of his fate at the end, when after murdering the Count he makes no attempt to escape and

*I am aware that there is a metaphysicalized view of Fascism and Nazism in particular that regards it as the most naked expression of human capacity for evil and tends to divorce it from historical and political circumstances; but it is a view I reject, if only because of its singular uselessness in attempting to predict and forestall a recrudescence of such ideologies.

calmly awaits arrest, look forward to the absurdist and existentialist currents that were to dominate French intellectual life after the war; yet nihilism notwithstanding his killing does liberate Garance from a relationship she has come to find oppressive, so that it is not merely a paradox to describe him as at once more negative and more positive than Gabin's incarnations of defeat.

That many films of the war years appear now in a profoundly ambiguous ideological light should not cause much surprise; to the difficulties of saying what one wanted to under Occupation must be added the often panic-stricken eagerness with which any degree of collaboration was later denied. Those film-makers who remained in Occupied France instead of emigrating to America as Clair and Renoir did often felt a need, economic as well as ethical, to disculpate themselves afterwards, which may account for the widespread attempts to recuperate films made under the constraints of Vichy by demonstrating that they were 'really' working against the system under the guise of compliance. Courtade's comments on *Le Ciel est à Vous* merit reproduction in their entirety:

> It is said and believed today that this film exalted the tenacity and the stubborn will-power of the people and that it was the silent cement of the Resistance. I was active in the Communist Party at the time, and I solemnly declare that it was perceived as propaganda for the class-enemy. Objectively, it served Pétain's interests, and the whole of the National Revolution [the Pétainist phrase for their attempt to rally France behind them] was to be found in it: the grumbling workman, well and truly one of us, with his glass of cheap wine, who taught industrial aviation a lesson; his wife, a pure heroine in the skies of glory, who had the humble virtues of our timeless soil; working like an ant to magnify one's dignity as a man . . . The D-Day landings were round the corner, the dictatorship was collapsing, and *Le Ciel est à Vous*, greeted by Vichy as a great nationalist work, was one of the last witnesses to the shameful 'Fascism with a French face' of Marshal Pétain.[13]

I leave these comments to stand, never having had the opportunity to see the film myself; but it appears plain that one of its main ideological thrusts, and one which must have benefited Pétain's efforts at national unity behind his prestige in the cause of collaboration, lay in its homogeneous presentation of the French nation. Gabin's typicality, in the Carné films, as a French working-man is, we have seen, undercut at several points, by his abandonment of his work and his sense of grievance and exploitation; this does not appear to be so in *Le Ciel est à Vous*. The exaltation of specifically 'national' virtues was a *locus classicus* of Fascism (perhaps one of the reasons why French Fascism never really got off the ground was precisely that, as Debray's remarks suggest, such qualitites had already been raised to the rank of universal virtues by the mythical importance of 1789). A film made under Occupation that presented these virtues as everywhere undiminished and as the cement of national unity was at least likely to arouse the hostility of anti-Pétainists.

Conversely, it could also be dangerous to construct too corrupt and divided a representation of French society, as Henri-Georges Clouzot found out when he attempted to resume his career after making *Le Corbeau*. This was actually a German-produced film, and thus not subject to Vichy censorship. It depicts life in a small provincial town as suppurating with guilts and jealousies that are brought to the surface by a spate of poison-pen letters. These turn out to be the work of a surgeon at the hospital (a morphine-addict to boot), who manages to get his wife committed to an asylum on suspicion of having written them but is stabbed by the avenging mother of one of his victims in the last reel . . *Le Corbeau* marks the first cinematic representation that I have been able to discover of provincial France as the site *par excellence* of hypocrisy and rancorous intrigue – an image indelibly associated nowadays with the films of Claude Chabrol. Its world is one in which everybody knows and lies to everybody else, but also one in which deceit and corruption go increasingly hand-in-hand with hypocrisy as one goes up the social scale. The film's image of France is anything but a united one, which might have been expected to make it unpopular with the Vichy régime; and Clouzot would certainly not have got away with references to drug-addiction and back-street abortion had the film been subject to normal censorship. But the Nazis in fact welcomed the film, as evidence of the decadence of the nation their puppets were busily 'civilizing' through the National Revolution movement, and Clouzot found obstacle after obstacle placed in the way of the resumption of his career after the war. This example taken together with that of *Le Ciel est à Vous* shows how the historical conjuncture can dramatically affect the way in which audiences are likely to derive an image of a society from a film, and also how films depicting French society as other than one united happy family were prone to draw the wrath of much of the public (especially the bourgeoisie) down upon themselves.

Jean Renoir

The films signed by Jean Renoir between 1935 and 1939 provide the best possible illustration of how the meanings of a film are not immutably present within it from the time of its making, but are constantly reread thereafter in the light of changing circumstances. This is why they will be considered only now, after our examination of French wartime cinema, for with the benefit of post-war 'hindsight' and an awareness of the historical and cinematic context that accompanied and followed them, their diagnostic value, and the ambiguities in which they abound, emerge devastatingly.

Le Crime de Monsieur Lange is described by Leo Braudy as 'the touchstone for the possibilities of community in Renoir's films of the second half of the 1930s',[14] and the courtyard in which almost all the action takes place, and round which the camera describes a 360° pan (as in *Red River*), has been seen as constituting an idealistic world of its own, within which after the defection of the scoundrelly publisher Batala the workers set up a cooperative to run their own firm. But this 'world of its own' is emphatically not a self-contained one, nor does it have the Arcadian elements of Clair's Paris. The cooperative is rooted in the economic need to survive as much as in the social need to reassert a community that has been vitiated by greed, financial (as when Charles is denied light by the advertising billboard nailed across his window) and sexual (Batala's blackmailing exploitation of his female employees). The precariousness of the cooperative is asserted as much as its community by the 360°

pan, for at the end of it Batala, believed dead, reappears from the shadows, and only Lange's shooting him prevents disaster. Lange and his girlfriend Florelle then flee towards the Belgian border, but are recognized in a café, where a popular court 'acquits' them after hearing the story.

A contemporary audience would have appreciated the immediate political relevance of the workers' cooperative, in the light of the Popular Front's policies, and perhaps also, depending on its degree of political sophistication, heeded the warning of the dangers of 'socialism in one courtyard' at the end . . . *M. Lange* is both an important intervention in the specific conjuncture of 1935-6 France and a broader statement about the need for class solidarity at a time when isolationist chauvinism was menacing peace in Europe. It is, in other words, both a French and a more 'universal' work, but in no sense an 'idealistic' one.

This becomes clearer when we reflect that it was immediately after shooting *M. Lange* that Renoir was asked by the French Communist Party to make a propaganda film for the Popular Front; the contemporary political implications of *M. Lange* were obviously not lost on them . . *La Vie est à Nous* was made for a tenth of what the average commercial feature film of the time cost, and was never screened in a commercial cinema before the war; it was shown to political and trade-union meetings and money to recover the cost of its making was collected at the end of the meeting, so that it can be said to represent one of the first attempts at an alternative political cinema through the conditions of its making and distribution as well as through its explicitly political subject-matter.

The film is a fictionalized documentary, denouncing the 'two hundred families' in whose hands most of the wealth of France was concentrated and the right-wing 'Croix du Feu' paramilitary group which had been behind the attempted Fascist coup of 1934. Its most interesting aspects from our point of view are the scenes in which France's national image and identity are discussed, with a view to ascertaining what a 'French style of socialism' might be – a problem that much later was to preoccupy the Popular Front's ill-fated successor, the *Programme Commun*, ratified in 1972. Schoolchildren at play discuss what the word 'France' evokes for them, and mention both its agricultural staples, such as wine and corn, and its luxury products for export (fashionable silk goods) or tourist consumption (châteaux). It is interesting to compare this with the reactionary agricultural populism denounced by Courtade in his analysis of Vichy cinema, and to see how one particularly pervasive perception of French national identity could be given diametrically opposed ideological inflections only a few years apart. Gaullism was to try to use in de Gaulle's own phrase 'a certain idea of France' to transcend supposedly 'outmoded' left/right ideological differences; it is unlikely that it could even have attempted to do this had the political ambiguity of French national identity – seed-bed for Gallic Fascism or necessary precondition for Socialism in France? – not been so thoroughly rehearsed in the political discourse and culture of the thirties.

The very places where *L'Humanité* (the Communist Party's daily newspaper) is shown being sold in the film are revelatory in this respect. It is sold in the suburbs of Paris (the traditional Communist stronghold), to bowls-players in the south (another region with a strong Communist tradition), and in a (presumably unidentified) port-town in northern France. The references to *pétanque* (the main working-class sport of the south) and to dock-labour anchor this part of the film firmly in proletarian culture and ways of earning a living, and would also have acted as a reminder of the French areas, geographical and cultural, in which the Communist Party was already well-established. Communism was then (as it is to a lesser extent now) equated by its nationalist opponents with the 'un-French' and the 'cosmopolitan' (the latter a camouflage for anti-semitic chauvinism), and the presentation of newspaper-sellers was clearly designed to combat such a perception.

La Vie est à Nous has remained a 'non-commercial' film to the present day. It was banned by the censors and shown only privately in France until 1969, and has rarely been available in Britain. The film's explicitly political provenance and subject-matter have thus effectively led to its exclusion from the Renoir auteurist canon. Attempts have been made to neutralize the socio-political import of his other work, but with a commissioned work of propaganda these would plainly not have been effective. Censorship, whether overt or covert (as through the non-availability of prints), was the only means left.

La Grande Illusion has frequently been vaunted as a timeless masterpiece of humanist cinema. This sounds as if I am mocking such claims, which deserve rather more serious analysis, and have indeed a measure of truth in them; the tension between concrete presentation and analysis of the conflicts within French society and an attempt to distil some immutable essence, not just of France but of Europe, is one of the major driving forces of the film. For Goebbels it was the most pernicious of French films, no doubt because of its favourable portrayal of a Jewish businessman (Rosenthal) and the plea for Franco-German unity implicit in the idyll between Jean Gabin/Maréchal and the German war-widow at the end. Left-wing groups greeted the film favourably on its release, yet the Belgian Socialist government banned it as too chauvinistic, and there are many aspects of the 'France' evoked in the film – an artificial, prison-bounded microcosm – that now appear ideologically highly ambiguous. This is partly the result of its implicit attempt to span the *entre-deux-guerres* period. Its proximity to the Second World War and treatment of Franco-German relations now cause it to be read as in some sense a 'document of the thirties', yet the First World War setting and references remain, so that it is not difficult to read the film as a generalized plea for Franco-German understanding, rather like Jean Giraudoux's *Siegfried et le Limousin*, which as both novel and play had been very influential in the previous decade.

The bond between cultures exists until the end only at the level of the aristocracy (the commander of the German prison-camp, Erich von Stroheim/von Rauffenstein, and the leader of the imprisoned French officers, Pierre Fresnay/de Boeldieu). This is largely for practical linguistic reasons, brought into relief when the French officers leaving their first prison are unable because of the language-barrier to tell their English successors that there is a partly dug escape-tunnel waiting for them. What it makes possible is an elegiac meditation on the impending decline of the aristocracy, whose thrust is deeply ambiguous. On the one hand, it appears prophetic of the large-scale breakdown in international communication that marked the late thirties in particular. On the other, it can be read as somewhat dismissive of the 'lower classes' (who in this film are not so low as all that – the action takes place in prison-camps reserved for officers, the inhabitants having been in civilian life variously actor, schoolteacher, engineer, and furrier, while the private soldier remains absent throughout). The figure of Rosenthal is extremely impor-

French Society through its Cinema

tant in this respect. Von Rauffenstein is dubious about trusting him; he is the butt of anti-semitic remarks which are not always neutralized by their context, and his wealth, mercantile not hereditary, which is expanding while de Boeldieu's is contracting, is the focus of a good deal of only partially jocular resentment. Rosenthal catalyses a double hostility (aristocratic and 'old' bourgeois) towards the 'new' bourgeoisie which made rapid fortunes from dealing in luxury goods, as well as an anti-semitism that, from the Dreyfus affair (1894) through to the Rue Copernic terrorist bombing of 1980, has never been far beneath the surface of French public life.

This emerges particularly clearly when Maréchal and Rosenthal are on the run together, after de Boeldieu has sacrificed himself, in a final nostalgic flourish of aristocratic style, to enable them to escape. Their exhausted quarrel, in which Gabin/Maréchal flings the coarsest racial insults at Rosenthal, is patched up in the interest of survival – a reconciliation which it might have been tempting to read, in 1937, as a lesson in the need for national unity (however conceived). But the projected final sequence of the film, in fact never shot, would have demolished such illusions. Rosenthal and Maréchal were to have arranged to meet in a Paris restaurant on the first New Year's Eve after the end of the war; the camera was to have shown a table with two empty places. Even without this final confirmation, the film as it stands provides sufficient evidence of the disunities and contradictions within French society to make its universal humanist aspirations (or the 'artistic brotherhood between nations that the film can play host to')[15] appear highly precarious.

The very title of *La Marseillaise* would have been seen as an appeal to national unity in the context of 1938. The history and ideological evolution of the song are a study in themselves; composed by Rouget de l'Isle as the battle-song of the revolutionary Army of the Rhine in 1792, it was taken up and sung by the Marseille Battalion on their march to Paris, which forms the narrative core of Renoir's film. Eugen Weber points out that this in itself was unusual, for most of the Marseillais and many of those through whose provinces they passed would not at the time have been French-speaking. The song thus favoured the Jacobin unification and centralization of France in two ways, through being the battle-anthem of the anti-monarchist armies and through helping the spread of French as the common language. In 1879 it became the national anthem, and by the beginning of the twentieth century its revolutionary aspects had been forgotton, subsumed into an all-purpose patriotism that enabled the right-wing nationalist author Maurice Barrès to write warmly of it in 1902.[16] In May 1968, it was the rallying-hymn of the Gaullist and right-wing counter-demonstrators, defending what they saw as the liberal-constitutional gains inherited from 1789. If the song appeared to have undergone an ideological *volte-face*, this was an illusion; it was rather that the significance of the 1789 revolution and the national unity it had supposedly forged had come to be radically reinterpreted.

In the pre-war and wartime periods (both First and Second), the Marseillaise was used as the signal for the suspension of ideological hostilities in face of a common enemy. It is defiantly sung by the prisoners in *La Grande Illusion* when Fort Douaumont is retaken (could Michael Curtiz have had this sequence in mind when he had Paul Henreid/Victor Laszlo drown out the Germans in *Casablanca* by leading the singing of it in Rick's Café?). In the Second World War, it was the anthem of the Free French, which is why it was so enthusiastically reprised by the Gaullists in 1968. The vicissitudes of the song indicate the ambiguities and problems attendant upon any attempt to rewrite the France of 1792 (or one of the Frances of 1792) into the context of 1938. For the supposed historical prototypes of the character in the film, for an audience of 1938, and for audiences since, the title would have evoked very different things – an indication of how equivocal the apparently simple notion of French national unity has been.

Nowhere does this appear more strikingly than in *La Règle du Jeu*. The first screenings of the film in the Paris of 1939 provoked bitter hostility and jeering (according to Renoir, who was in the projection-booth, it made no difference if contentious sections were omitted; every audience found something new to jeer). After three weeks, the film was withdrawn. In October 1939, the government banned it. The Germans followed suit during the Occupation. The original negative was destroyed by a bombing raid during the war, and in 1946 only mutilated versions were available. It was not until 1959, at the Venice Film Festival, that the film was finally shown in its entirety (almost – there are still certain sections for which the soundtrack alone survives).

So chequered a history has more than curiosity-value. There remains in *La Règle du Jeu*, as there has been ever since its first screening, something to antagonize any spectator tempted to view French society as a unified national whole. The film's intrigue appears to pit the 'old' France – landed, affluent, attached to a liberalized version of the old noble traditions – against the 'new' – individualistic and entrepreneurial (whether in the person of the record-breaking aviator André Jurieu or in that, more literally down-to-earth but no less irresponsibly Bohemian, of the poacher-turned-servant Julien Carette/Marceau). In between the two, the 'conventional' bourgeoisie – the world of teachers, engineers, and civil servants whose denizens fill the prison-camps of *La Grande Illusion* – is conspicuously absent. Could it have been this daring eviction of the class which since 1789 has considered itself the backbone of European civilized progress – and which would certainly have provided the bulk of the film's 1939 audience – that so antagonized its spectators?

The film's credits propose a division of characters into 'masters' and 'servants' that thenceforth works only to undercut itself. Of the 'masters', the Marquis de la Chesnaye is Jewish (which is pejoratively alluded to in the film by representatives of both classes); his wife Christine is Austrian (Nora Gregor's German accent sparked off the first signs of audience discontent at the 1939 screenings); André Jurieu is an aviator, described by Octave as 'not capable of crossing the Champs-Élysées on foot without a pedestrian crossing',[17] and an obvious prototype of the Bohemian freelance (he is interviewed on the radio at the beginning, thereby linking himself with the up-and-coming but still untrustworthy world of what was later to be called the 'media'); Octave is referred to as 'a dangerous poet',[18] and alludes disparagingly to his own financial (and, we may infer, emotional) parasitism. The 'servants' are led by the jealous gamekeeper Schumacher, obsessively conscious of his place in the estate's hierarchy, yet the character most disruptive of that very hierarchy when he chases his wife's aspirant lover with a gun, transforming his passions into reality without recourse to the mediations of social convention. It would also have been relevant to a 1939 audience that he is from Alsace – a border-province then, as often before, claimed by Germany. The servants also include Corneille, the majordomo, named after the dramatist who came to symbolize for generations of French students the noble conflict between love and duty, and the poacher-turned-servant Marceau, who accepts his banishment from the château at the end as tranquilly as he accepted his invitation to join its ranks at the beginning.

The servants' world, in other words, includes a disruptive representative of 'Teutonic' passion erupting though the veneer of Gallic style, flanked by representatives of the worlds of high tragedy and low farce. That of the masters is (culturally or racially) 'adulterated' by its own stultifying standards in its principal representatives, while among the supporting cast we find a homosexual and a South American. The very order articulated and set up in the presentation of the film's cast is worked through and gnawed away by a tissue of contradictions and incongruities, whose impact grows in proportion as the spectator is able to place himself in the position of the uneasily xenophobic French bourgeoisie and nobility of 1939 and to realize how each member of the *dramatis personae* operates a subversion of his own (and any audience's) position.

It is even more appropriate than it may seen that the film's last word should fall to the character designated simply as 'the General', whose handlebar moustache and would-be sly asides have gone to mark him out as the representative of the old gentlemanly France. After Jurieu has been shot, the Marquis de la Chesnaye pronounces an elegiac speech from the steps of the château, enjoining his guests to mourn 'this wonderful friend, this excellent companion'. Saint-Aubin, the old order's most insensitive representative, comments acerbically: 'A new definition of the word ACCIDENT!' – with some justice, considering that the Marquis could have asked for nothing better than to be rid of Jurieu's attentions to his wife. Fortunately, however, the old order is – for the moment – still capable of going beyond such an admission of defeat; the General ripostes: 'No, no, no, no, no! La Chesnaye does not lack class, and that is a rare thing, these days, my dear Saint-Aubin, believe me, that is a rare thing!'[19]

This closes the film's narrative, and its representation of French society, in a number of ways. It marks the final admission of La Chesnaye, hitherto often belittled as a Jewish parvenu, to the ranks of the 'true' nobility – hence, paradoxically, the beginning of the end for that nobility by its own xenophobic standards. It stitches together a conspiracy of silence which ensures that the participants in the ballet of relationships that has gone on throughout the film all return to their original places: Marceau goes back to poaching, Schumacher and the Marquis remain, for better or worse, with their wives, Octave returns to the Bohemian void whence he came. It thereby presages the collapse of the old order (implicit from the beginning of the film in the subversion of the master/servant taxonomy) far more devastatingly than Saint-Aubin's remark could ever have done. Not so much a conspiracy of silence as a conspiracy to take events at their face value – to disregard the gaping holes opened within the social order by the weekend's events – represents the last gasp of the gentlemanly world within which a tacit understanding based on a shared style was enough to transform conflict into acquiescence. In much the same way, in Beaumarchais's play *Le Mariage de Figaro* which is quoted in the film's credits, the Count (representative of the pre-1789 dominance of the nobility) is allowed to retain the illusion of triumph, but it is plain that his empire is crumbling. When the guests return to the château in the Renoir film, it is as though for the last time, in the knowledge that everything is still the same, but that nothing will ever be quite the same again. Less than two months after the première, war had broken out.

It is instructive to compare the ending of *La Règle du Jeu* with that of *Citizen Kane*. Both films know something that the characters within them do not (Kane's secret, the impending collapse of the old France), and both endings have been given intensified force by events subsequent to the making of the films (Welles's blighted career, the onset of war). But the type of knowledge is radically different. In *Kane*, it takes the form of populist revelation by the camera as omniscient narrator, in *La Règle du Jeu* of an uneasy harmony 'neutrally' observed and disturbed only by Jurieu's death and, more subtly, by the defeated departure of Octave – played by the film's director . . . The difference speaks eloquently of the distinction between European and American styles and conventions of direction, and, more significantly for our purposes, of the gulf that separated American populism, based on charismatic razzamatazz even to its final paradoxical discomfiture, from European aristocracy, a stylish façade refusing to acknowledge that the beams supporting it were rotten.

The New Wave

The post-Liberation cinema will occupy relatively little of our space, for it laboured under a twofold constraint – economic and artistic. Many of the best directors and performers were either blighted by suspicion of collaboration or pursuing their careers in Hollywood, and the combination of large-scale imports from the United States, technical innovations such as the wider screen which the undercapitalized French industry could not hope to compete with, and an inadequate quota-system for the showing of foreign films meant that the industry was able to reassert itself only with the aid of governmental subsidies and large-scale international co-production.

This need not have mattered too much (though the scope French society had to represent itself through the cinema would necessarily have been restricted by internationalism and co-production, whatever their other benefits), but it was accompanied, especially during the fifties, by a sinister intensification of censorship. In the United States and Britain, this has tended (with the obvious exception, perhaps so overwhelming as to disqualify the point, of the HUAC years) to be 'morally' rather than politically articulated. In France, where political and ideological discussion and struggle has at least since 1789 occupied a more important place, overtly political censorship has been the rule rather than the exception. René Vautier was sentenced to a year's imprisonment in 1955 for his anti-colonialist film *Afrique 50: Le Rendez-Vous des Quais*, a documentary about Marseille dockers demonstrating against the colonial war in Indochina, which was promptly banned. The most notorious example of direct governmental censorship since the war – the banning of *La Religieuse*, set in the eighteenth century – is in some ways the least typical, if only because it was so manifestly ludicrous that within less than two years after the film's banning in 1966 it was being widely screened and achieved an unsurprising *succès de scandale*. More discreetly obnoxious were the system of local censorship, which allowed mayors to prohibit in their towns films approved by the national censors, and the liberal use of the power of veto to suppress cinematic criticism of the eventually disastrous colonial ventures in South-East Asia and North Africa. When one bears in mind that ORTF (the national radio and television authority) has, despite the courageous efforts of many of its staff, always played the role of ventriloquist's dummy to the government, the lack of channels for filmic presentation of criticisms of, or alternatives to, governmental policies becomes dauntingly apparent.

By the mid-fifties, the cinema had dwindled to become only the seventy-sixth industry in France, and the French public went to the cinema less often than their Spanish or British counterparts. The absence from their screens of any remotely adequate treatment of the crises that shook the country is likely to have been partially responsible for this. 1954 marked the beginning of the Algerian movement towards independence – a conflict that was to drag on until 1962 – and the end of the calamitous adventure in Indochina. Little more than three months – the span of a university summer vacation – separated the termination of one ill-fated colonialist enterprise from the beginning of another; yet it was not until 1957 that the first French feature film set in Indochina – *Patrouille de Choc* – was shown, and *Le Petit Soldat*, completed in 1960, was banned until after the end of the Algerian war because of its frank references to the use of torture. The sixteen-year death-agony of a major colonial empire was evicted from its national cinema, through the silence of many of its producers and directors and the outright repression of the rest.

The last years of the fifties marked a renaissance, from what was initially a quite unexpected quarter. A number of young directors – some critics, some graduates of the IDHEC film-school – began to make independent low-budget films, lumped together under the misleading title of the 'New Wave'. From our point of view there are two major objections to this label. It homogenizes a range of directors whose social and political attitudes differed widely, and it privileges – understandably but unduly – the techniques of their filming (location-shooting, hand-held cameras, later direct sound) over its locales and the milieux it evoked and reproduced.

Thus, the 'New Wave' is usually seen as arrogantly Parisian, and operating within a very restricted conception of Paris at that (the Latin Quarter, Montparnasse, and the area round the Champs-Élysées). There is some measure of truth in this; Paris, being the world capital of cinematic discussion and study, attracted aspirant directors like a magnet, and it was natural that those who had studied or writtten about cinema there should also make their directorial débuts on location in the crowded streets and boulevards. But it is only part of the story. Godard's lovers in *A Bout de Souffle* stage their romance in a Paris of boulevards, ever-open cafés, and a business and financial sophistication that frustrates Jean-Paul Belmondo/Michael Poiccard's attempts to secure his criminal profits (which come in the form not of banknotes, but of a cheque he finds it impossible to cash). But at the same time he dreams only of escape – culturally via American movies, geographically to the Provençal coast, significant both as the traditional Mecca of sun-seekers and because its capital, Marseille, is the focus of much large-scale organized crime. And Jean Seberg/Patricia Franchini is an American, half-studying in Paris and treating it as a kind of distillation of the whole of European culture. Their adventure is thus *staged* in a Paris to which neither fully belongs. One can say that the film is certainly Parisian, but its characters are not. Beyond the bars and offices where Michel attempts to secure his loot, or the Champs-Élysées where Patricia sells her American newspapers, glimpses of many other worlds are periodically vouchsafed.

It would be possible to turn this argument on its head, to say that the very Parisian-ness of 'New Wave' Paris lies in its cosmopolitanism and anonymity, in the fact that (quite unlike the solid rural France extolled by Vichy) it is a world that flaunts its rootlessness as a badge of pride. On this reading the nervous improvisations of *A Bout de Souffle* or *Bande à Part* would be the cinematic correlative of Paris as capital of the Existentialists, just as the world of René Clair corresponds to the city's earlier incarnation of *la vie de Bohème*. But it remains true that many of the best-known films made by young directors in the fifties and sixties have a focus outside Paris, however 'metropolitan' in the wider sense their style of filming may be. Truffaut set much of *Tirez sur le Pianiste* in the French Alps, and broadened out *Jules et Jim* from its Parisian beginning to take in the tortured world of First World War Europe, before the tragicomic ending in the countryside. *Jules et Jim* can indeed be seen as an elegy for Bohemian Paris: the two friends of the title, who meet while preparing for a costume-ball and are shown at the beginning sharing a number of artistic and sporting interests, are almost torn apart by the twofold strain of war between their countries (France and Germany) and their rivalry for the love of

Jeanne Moreau/Catherine. Catherine's capriciousness and unpredictability can then be read as the extreme manifestation of that Bohemian spirit – romantically heedless of boundaries and restrictions – which received its *coup de grâce* between the First World War and the Depression.

One director in particular, as has already been mentioned, made the pettiness of provincial life his speciality. Chabrol's *Les Bonnes Femmes* is, admittedly, set in Paris; but it is a Paris a long way removed from the boulevard bustle of *A Bout de Souffle* or the romantic comradeship of *Jules et Jim*. The four shopgirls misogynistically referred to in the title are like variations on a theme by Emma Bovary, their dreams of escape are snobbish or tawdry, and the only exception (Jacqueline) is taken to the countryside by a motorcyclist near the end, apparently for a romantic idyll but in fact to be brutally murdered. The Paris of this film differs from provincial towns only in that its nastiness and virulence are on a grander scale.

This is not to suggest that Chabrol's view of the French provinces is an entirely uniform one. Saint-Tropez, Riviera haunt of the wealthy and hedonistic, is the setting for the sex/power/death triangle of *Les Biches*. The homicidal urges of Charles in *Que la Bête Meure!* dedicated to avenging his son's death at the hands of a hit-and-run driver, are mollified by the growing closeness between him and the killer's own son, but also by the very setting of the film. At the end, probably to spare the son any suspicion of having killed the man they both loathe, Charles puts out to sea in his yacht, to the accompaniment of Brahms's *Four Serious Songs*, based on a text from Ecclesiastes. This uneasy amalgam of Christian self-sacrifice and pantheistic absorption into the infinite has been prefigured earlier in the film, particularly in the opening sequence (which shows Charles's son about to be run over, with the same music as background), but also through Chabrol's use of the Breton land- and seascape. Brittany is the most devoutly Catholic region of France (shrines and crucifixes abound beside its roads, as they do in the film), and the one with the longest and most rugged coastline – with, however unsatisfactory of definition the term may be, the most mystical landscape. Its use in this film thus intensifies the ambiguous spiritual resonances as no other province could have done.

The central region of France is the setting for *Les Noces*

Rouges – suppressed by the government until after the 1973 elections because it allegedly made the Gaullists look ridiculous, and based on a real-life triangle of jealousy and murder. This had occurred in the small town of Bourganeuf; Chabrol transposed the intrigue to Valençay, in the adjoining *département*, and extracted maximum mileage from the soporific countryside surroundings and ingrained social conservatism, against which the wheeling and dealing of the Gaullist mayor (murdered by his wife and her lover, and clearly designated as impotent) provides an even more unpalatable reaction.

Chabrol did set two feature films, *Les Cousins* and *Les Godelureaux*, in the intellectual milieu of Saint-Germain-des-Prés; but the bulk of his output deals with the provinces, albeit through the eyes of a particularly jaundiced Parisian. Alain Resnais exploits the resonances of different provincial towns in *Hiroshima Mon Amour* and *Muriel*, the first set in Japan but flashing back to events in wartime Nevers, in the heart of Occupied France, the second in Boulogne-sur-Mer, destroyed in the war and reconstructed in a rather confusing style of which much is made in the film. Jacques Tati – not a 'New Wave' director in even the most elastic sense of that overused term, but a key influence on the evolution of cinematic language in France – set *Jour de Fête* in a village in central France and *Les Vacances de M. Hulot* in a Normandy seaside resort where the central character is on holiday. And in Louis Malle's *Les Amants*, whose supposed erotic frankness and unfashionable romanticism have tended to block other aspects of the film, the tension between the high society of Paris and that of the provinces is an important dramatic element, ambiguously resolved at the end.

Les Amants shows Jeanne Moreau/Jeanne Tournier living in a world that could almost be described as a passionless bourgeois updating of that of *La Règle du Jeu*. Her husband owns a newspaper in Dijon, the capital of Burgundy; he is so preoccupied with his work and the social and business contacts it procures him that their eight-year marriage is clearly no more than a shell. She has a lover, who plays polo in the Bois de Boulogne – a park on the outskirts of Paris, between the sixteenth *arrondissement* and the suburb of Neuilly-sur-Seine. This would have been important to a French audience as the locale of a whole haut-bourgeois life-style (an equivalent area in London would be

Knightsbridge or Belgravia). Her regular weekend trips to Paris to meet her lover are camouflaged as shopping and social occasions with a female friend, but this is less a guilty alibi than a white lie of convenience; so detached is her husband that were she to come in and announce that she had just returned from an amorous assignation, he would, one suspects, barely notice the difference. The world of Dijon and that of Paris, far smarter though the latter is, have in common an obsession with material possessions and a total lack of spontaneity or passion. The polarization of the world of provincial Burgundy and that of smart-set Parisian intrigue recalls the literary work of Colette – a Burgundian profoundly attached to her native province, but also a denizen of the Bois de Boulogne world so vituperatively criticized in her work.

The student with whom Jeanne escapes the stifling constraints of her two lives is likewise in a well-established French literary tradition – that of the younger man being initiated (emotionally if not physically) by the older woman (examples include Emma and Léon in Flaubert's *Madame Bovary*, and Léa de Lonval with Fred Paloux in Colette's *Chéri*). The film's ending has been much criticized for its supposed implausibility – Jeanne leaves husband, lover, and child to drive off with Bernard we know not whither, after one delirious night of love – but perhaps precisely such implausibility was needed to emphasize the deadening tyranny of routine. The soaring romanticism of the night-time scenes acquires its force largely from the tedium of day-time life, and the absence of any definite goal for the lovers at the end can be seen as a reaction against the impeccable geographical and chronological order of Jeanne's life hitherto. The film is first of all a denunciation of both Paris and the provinces and only secondly a love-story teetering on the brink of excess.

Godard shot *Le Petit Soldat* in Geneva – maybe as a conscious homage to Joseph Conrad whose *Under Western Eyes* also takes place in that city, maybe in reference to Switzerland's long-standing policy of diplomatic neutrality. For the film conspicuously fails to address itself to the rights and wrongs of the Algerian war that is its ostensible subject matter. The central character, Michel Subor/Bruno, is working for a Fascist terrorist organization against the Algerian freedom fighters of the FLN, to which his girlfriend (Anna Karina/Véronique) is subsequently revealed as belonging. But the two sides are not constructed as in any way fundamentally different, which caused Godard to be reviled as a neo-Fascist by many on the left and also led to the banning of the film for three years because of its supposed pro-nationalist slant.

It is easier to see now, twenty years later, how Godard's concern with the interrelationship of the political and the personal, with what it means to conceive of and carry out a political act, is present and at work in *Le Petit Soldat* behind the absence of more obvious political considerations. Easier, too, to realize the importance of the reference to torture (here as in Resnais's *Muriel*, which appeared in the same year), for it was revelations about the widespread use of torture by the French troops in Algeria that first alerted French public opinion to the whole question of the country's status. Dislike of Arabs, perhaps even more prevalent in France than anti-semitism, is exemplified in *Muriel* through the comments of Alphonse, surely the definitive café bore: 'I respect all races even if I can't stick Arabs.'[20] Both films also mime in their very subject-matter and structure the silence and falsification of the media while the war was going on. *Le Petit Soldat*'s apparent neutrality, of both treatment and setting, and the home movie of Algeria obsessively screened and rescreened by Bernard in *Muriel* – an insignificant and ineptly shot travelogue with no indication even that it was shot in wartime – both demonstrate how difficult it was to construct a picture of the war in any way critical of the colonialist policy.

That two Godard-signed films figure among those banned or otherwise censored will cause little surprise to those who know him as the film-maker – probably the artist in any domain – whose work most consistently criticizes the social practices and underlying assumptions of contemporary France. The fact that the two films in question date from well before his explicitly 'political' period emphasizes how concerned he has always been with questions of (usually but not always French) social reality, its construction and representation. 1965 can be seen as a major political turning-point for him. By inflecting the science-fiction genre towards a concern with the politics of language in *Alphaville*, he gave his long-standing obsession with different cinematic forms a more overt social bias; by shooting the film entirely on real Paris locations at night, he revealed the subversive non-realist possibilities of location filming, and made the fictional allegory's contemporary relevance plain through the combination of familiarity and strangeness in the settings. *Pierrot le Fou*, complement to *Alphaville* in so many ways, contrasts Riviera light to the earlier film's Parisian darkness. Watching Jean-Paul Belmondo/Ferdinand Griffon/Pierrot le Fou escape the centralized constriction of Paris, one is reminded of Godard's own departure to make video films in Grenoble (in the early seventies). Both central character and film-maker developed new expressive and linguistic forms (whether experimental journal or political video) away from the domination of the capital.

Pierrot le Fou castigates fashionable Paris in the party-scene at the beginning, whose polite clichés are interrupted when Pierrot flings a cake at the guests; but it also subverts the traditional view of the Mediterranean as joyful suntrap, and of the 'simple life' as that most conducive to creativity. The landscape as he and Anna Karina/Marianne drive south is menacing and death-laden, with its overbright glare and the crickets twittering like a threat; and the tranquil refuge of the artist is interrupted by the connected concerns of love and politics. Gun-running and Vietnam provide first a background, then a foreground, of violence – perhaps the return of the political violence repressed or censored in Godard's earlier work. One brief shot just after Pierrot and Marianne have arrived on the coast shows Pierrot driving a tractor, which appears nowhere else in the film and whose presence is never accounted for. It functions as an ironic glimpse of the 'drop-out' existence for which so many educated French and others were to yearn in the late sixties.

Godard's own involvement with the problems of contemporary France became considerably more marked in this period. In *Deux ou Trois Choses que je sais d'elle*, he resumes the concern of *Vivre sa Vie* with prostitution; only here Marina Vlady/Juliette Jeanson prostitutes herself for luxury goods, not for the bare necessities of survival, and the film's preoccupation with advertising, tower-blocks, and the continuing horror of Vietnam constitutes a powerful criticism of the wholesale Americanization of France. De Gaulle's inclination in foreign policy had always been to seek his allies in Europe rather than across the Atlantic, but the phenomena portrayed by Godard show that this policy, however successful on its own terms, manifestly failed to prevent large-scale cultural imperialism.

La Chinoise actually takes a political group as its subject-matter; the Maoist group on whom it centres moves during the course of the film from communal discussion and action to a split involving the expulsion of one of their members and the (abortive) use of guerrilla tactics. What is interesting about the film from our point of view is its extraordinary prescience. Disillusionment with, on the one hand, the poor quality of social life in the Soviet Union and, on the other, the French Communist Party's loss of contact with new modes of thought and activity caused many on the French left in the late sixties to turn towards the example of China – the China of Mao Tse-tung, at once poet, philosopher, and revolutionary,* and that of the Cultural Revolution. To quote Sylvia Harvey:

> For it is clear that the revival of interest in problems of ideology in France was in certain respects fostered by an encounter with the ideas of the Chinese Communists who were emphasising . . . the importance of a struggle consciously conducted by the proletariat within the sphere of culture and cultural production even *after* the taking of power by a proletarian party and the revolutionising of the ideas of production.[21]

This was to help, in May 1968, what started out as a piecemeal array of strikes by groups of students and workers to develop into a major political upheaval, and also an important forum of cultural struggle. Many of the arguments that raged during that period (especially between the 'Establishment' left, led by the Communist Party, and the *gauchistes* of far-left or anarchist inclination) are strikingly rehearsed in *La Chinoise* when one of the members of the Maoist group has a conversation on a train with her old philosophy lecturer – Henri Jeanson (playing himself), a prominent member of the Communist Party. Their argument about the merits of spontaneous action and large-scale mass struggle was to be echoed (though in neither so coherent nor so amicable a way) throughout the debates of May 1968.

More striking still to a contemporary audience is the monologue of the expelled member of the group at the end of the film.

---

*The hagiographic tone of this description intentionally corresponds to the deification of Mao that became prevalent on the European far-left in the late sixties and early seventies.

He expects to return to Besançon, the provincial city from which he came to Paris, and thinks he may well (re)join the Communist Party. If the conversation on the train is uncannily prophetic of French left-wing politics of the late sixties, this monologue is no less so of what was to happen thereafter. The failure (or at best partial success) of May 1968, whose would-be revolutionary torrent trickled away into a tranquil reformist brook, led many fired by the impetus of the Parisian student movement to change their direction. Many went to Trotskyist or ecologist groups, or left politics altogether for jobs, mysticism, or a combination of the two; but there was also a steady flow of erstwhile *gauchistes* into the parties of the 'Establishment' left. Whatever our judgement on the expulsee, he does today represent an identifiable *political* type. The open structure of *La Chinoise* (which refrains from any omniscient pronouncement on the positions of its characters) ensures that his decision appears as historically determined, and subject to the verdict of history. After the collapse of the left-wing *Programme Commun* in the late seventies and the return of the Communist Party's old subservience to Moscow (manifested in a less-than-enthusiastic condemnation of the Soviet intervention in Afghanistan), the pendulum now appears to be swinging in a different direction. How this discourse will appear in five years' time, after a period of Socialist rule, or what kind of discourse might have taken its place, remains to be seen. The triumph of Godard's film is that it makes the posing of such questions possible.

The materialistic Americanization of French society, the Parisian colonization and despoiling of much of the countryside near the capital, and the sometimes grotesquely exaggerated forms of political action that sprang up in the late sixties fuse in *Weekend*. The married couple played by Jean Yanne and Mireille Darc (stars from more popular French cinema) are held together only by their desire to murder the wife's mother for her fortune, and by their homicidal designs on each other. The Norman village where the mother lives (a real place) is called Oinville ('*oin*' in French means the 'oink' of a pig), and allusions to greed, piggishness and cannibalism abound, culminating in the wife's final eating of a meal consisting partly of her husband in the company of the guerrillas of the 'Seine-et-Oise Liberation Front'.

Deux ou Trois Choses, *La Chinoise*, and *Weekend* constitute a remarkable introductory triptych to the 'events' of May 1968. Films dealing with the aftermath of these, and where the 'children of 1968' are nowadays to be found, are beginning to be made (an example is *Une Semaine de Vacances*), but the primary interest of the 'events' for us is in the way in which the cinema industry was deeply involved with them from an early stage. The media (particularly ORTF) were obviously involved from the beginning – the governmental constraints on broadcasting and the question of what form coverage of the events should take saw to that – but the involvement of the cinema requires some explanation. It arose partly because of the dismissal of Henri Langlois, head of the Paris Cinémathèque, in February – an iniquitous decision rescinded in April, but not before the film world too had mobilized its indignation at government interference in its business. During the 'events', uncensored documentary and agitprop films were made and (illegally) shown; the 'Estates General of the Cinema' convened technicians, directors, and others active in French film to discuss a variety of projects for reforming or revolutionizing the area of cultural production (especially in the cinema); and the Cannes Film Festival, France's leading cinematic showcase and an often philistine bourgeois 'cultural' jamboree, was cancelled.

The record is a proud one, but its effect on the films made thereafter was conjectural. *Cahiers du Cinéma* and *Cinéthique*, probably the two main French magazines of cinematic theory and criticism, moved towards a complex revaluation of their position, characterized (to oversimplify) by conjugation of new Marxist notions about the importance of cultural struggle with ideas on subjectivity and identification drawn from a rereading of Freudian psychoanalysis.* The impact of these developments on film theory has been immense in Britain and the United States. Within the French cinema, for reasons of economic pressure and revulsion from existing institutions alike, it has been considerably less.

Godard, in whose work their effect is perhaps most clearly perceptible, was able to achieve this only by a dual process of withdrawal – from Paris to first Grenoble and later Switzerland, and from the world of 'proper' film-making and distribution to

---

**May 68 and Film Culture* is a good summary of these developments in their relation to the events of May.

independent experimentation with new forms, particularly video. Of his seventies work, *Numéro Deux* has so far proved among the most widely available in Britain. Made on video in Grenoble, in collaboration with Anne-Marie Miéville, it retains many of the thematic concerns of his earlier work – the debilitating effect of life in a high-rise estate, the connection, carefully occluded by bourgeois society, between marriage and prostitution, the closely allied connection between sex and politics – but articulates them in a number of very different ways. The theme of blockage and malfunctioning assumes a new importance; the wife is constipated, her husband impotent, the lavatory permanently blocked – all so many indications of how the constant process of passive consumption, already a significant factor in *Deux ou Trois Choses*, can work destructively upon the individual bodies that go to make up society. The use of video and split-screen allows Godard to juxtapose these concerns. *Numéro Deux* provides a stimulating illustration of how a new type of filming, outside both the financial constraints of the standard distribution circuits and the narrative constraints of conventional plot, can tackle the problem of contemporary France in a different way.*

We have seen how key moments in French history (such as the Revolution of 1789 and the First World War) were re-articulated in its cinema at later periods to which they appeared particularly relevant. The myth of omnipresent resistance to the German Occupation came in for some stringent scrutiny in the seventies, initially via *Le Chagrin et la Pitié*, a lengthy documentary montage about Clermont-Ferrand under Occupation which suggested how widespread collaboration had been. The reasons why it was at this time that certain unpalatable truths of wartime appeared on the cinematic agenda are not so clear as they were for *La Marseillaise* and *La Grande Illusion*. Likely contributing factors were a sense of impasse among film-makers on the left after the failure of May 1968 and the uneasy retrenchment of the *Programme Commun*, and the sense that the Gaullist consensus that had dominated power in France since 1958 was reaching its end. President Pompidou's death in 1974 and the election of Giscard d'Estaing in the same year went to confirm this feeling.

---

*See *Godard: Images, Sounds, Politics* for a fuller survey of Godard's seventies work.

Gaullism had predicated itself on the heroism of the Resistance; once it had been seen to have run its course, certain cherished myths became much easier to question.

Lacombe Lucien marks the major feature-film treatment of the problem of collaboration. Set in the south-west, it presents its central character in a rural dead-end milieu (slopping out in an old folks' home, pushed out from the family house by 'friends' of his mother's lover, finding his only consolation in poaching), from which he initially seeks to escape by joining the Resistance. Refused as too young and irresponsible, he promptly justifies this rejection by transferring his attentions to the occupying forces, to whom he betrays the name of the local Resistance leader – his old schoolmaster, the agent of his rejection. At no point in the film, not even when his affection for the daughter of a Jewish tailor whom he helps to escape conflicts with his supposed 'duty' to the anti-semitic cause, does Lucien appear remotely aware of any wider implications to his action. This is at once the film's strength and its weakness – its strength, because it demonstrates how involvement with one or other side of the struggle could fulfil a similar role to playing sport or joining the army, as a means of injecting desperately-needed excitement into a tedious existence; its weakness, because the result and effect of collaboration upon the collaborators are imperfectly illustrated by a character so obtuse as Lucien. Reactions to the film varied widely; some saw it as telling unpalatable but necessary truths, others as blackening the fair name of Resistance France, others again as reactionary in its very attempt to bring its audience close to the motives and identity of a collaborator. The malaise which the film both portrayed and produced reflects as much upon the mood of (soon-to-be) post-Gaullist France as it does upon what may 'really' have happened in 1944.

Français, si vous saviez . . ., whose makers had also worked on *Le Chagrin et la Pitié*, is a more direct testimony to how various sectors of French society adjusted to the demise of Gaullism. The film (shown in three long sections) operates through a montage of interviews and documentary footage to puncture the Gaullist legend, but as Marc Ferro has pointed out it does so in a more questionable way than might be supposed. To quote:

This film settles grievances with de Gaulle; to do this efficiently, the makers *called upon all layers of anti-Gaullism*: victims of the Liberation, disappointed members of the Resistance, people repatriated from Algeria, and so on. So clearly angled are the questions that even Soustelle, hardly likely to be well-disposed towards de Gaulle, finally takes up his defence.[22]

What is significant from our point of view is the way in which the film seeks to construct a populism that is the mirror-image of de Gaulle's own. The final footage of the film's third part, showing de Gaulle's burial-place at Colombey festooned with Lorraine crosses and ex-voto-like messages from admirers who clearly literally expected a rapid resurrection, is a frightening denunciation of popular gullibility; but the film's title, and its explicit addressing of the very 'Frenchmen and Frenchwomen' to whom de Gaulle invariably referred at the beginning of his speeches, indicates that Harris and Sédouy, here at least, construct a homogeneous view of the 'French nation' that is nowhere clearly differentiated from de Gaulle's own. Ferro says:

> . . . de Gaulle lies, he has always lied, that is the conclusion implicit in the whole of the film; he lies like all politicians . . . for Truth lies in structures, in permanence: the choice of a traditional province such as Lorraine at the beginning of the film clearly denotes its 'anti-political' – or 'anti-politician' – stance.[23]

Which returns us once more to the observations of Debray quoted at the outset. The recurrence in the 'commercial' cinema of events (or pseudo-events) from relatively recent French history, and the alternate questioning and reassertion of the unity of the French people, emphasize a sense of privileged nationhood and a preoccupation with one's own national history, its great moments and inglorious connecting-tissue, that go well beyond the specifically cinematic and are rooted in the whole complex network of French cultural and ideological life. Be it Renoir rewriting 1789 into 1938, Godard and Resnais alluding to an Algerian conflict it was impossible to tackle more overtly, or

Harris and Sédouy implicitly setting up a posthumous 'Popular Front' (but an ideologically 'neutral' one) against General de Gaulle, the representation of French society in its films is largely dominated by the relationship between its history and its – shifting – present.

Part Three

'You will, Oscar, you will' – British Disavowal and Repression

In international affairs – a reputation for disinterestedness
In national affairs – a tradition of justice, law and order
In national character – a reputation for coolness
In commerce – a reputation for fair dealing
In manufacture – a reputation for quality
In sport – a reputation for fair play.[1]

> Sir Stephen Tallents listing some suitable subjects
> for films to be produced by the 1930s Empire Marketing Board

My mother-in-law is like a beautiful piece of antique furniture – a large chest, with drawers!

> Well-known British comedian, as part of his cabaret
> at a 1968 Oxford Commemoration Ball

Oscar Wilde (*à propos a particularly witty remark*):
   I wish I'd said that.
James Whistler: You will, Oscar, you will.

Three quotations at first sight wildly divergent, bound together only by the unmistakable 'Britishness' of the first two at least, and the clear British provenance of the third. But between them, and across the seventy-odd years that span them, a great deal is said about the culture of Britain, about not only what it flaunts but also what it represses and partially disavows. I had better preempt one important criticism here and admit that for the purposes of this brief study 'Britain' will effectively mean 'England', and usually though not always metropolitan England. This is the result partly of my lack of space here, partly of the marginalization of most of England north of Watford (except, of course, for the 'kitchen-sink' films of the fifties) and the virtual exclusion of Wales and Scotland from the dominant cinematic culture. Thus the pawky eccentricities of *Whisky Galore* and the sombre tableau of deprivation – cultural and emotional as well as material – painted by Bill Douglas in his trilogy (*My Childhood/My Ain Folk/My Way Home*) represent two polar views of 'Scotland' between which remarkably little connecting-tissue is to be found. The views of Britain constructed in its cinema have been overwhelmingly English-based.

Tallents's litany, today almost comic in its stiff-upper-lip self-righteousness, is most certainly English rather than British, and unquestionably metropolitan English at that. (I use the term 'metropolitan' rather than 'London-centred' simply because the virtues Tallents adduces, and the unhesitating acceptance of them, are now, as they have probably been since the thirties, more firmly rooted in the London suburbs and the 'stockbroker-belt' and 'Costa-Geriatrica' outposts than in the capital city itself.) The characters in British cinema who distil the qualities he vaunts most clearly, and with a sense of humour itself all too 'British', are the two clubmen played by Basil Radford and Naunton Wayne – Caldicott and Charters. In *The Lady Vanishes*, set in Central Europe, their running obsession is with the latest score in the Test Match back home; the fact that this is taking place in Manchester will serve to alert most metropolitan British audiences to the inevitable conclusion ('Floods in Manchester – Test Rained Off'). Caldicott's calm under enemy fire at the end, even when painfully wounded, and the two men's scornful attitude towards the undemocratic régime of the anonymous Central European country in which they find themselves further go to reinforce their 'Britishness' *à la* Tallents. Dame May Whitty/Miss Froy, who manages to escape with the vital coded message, and whose tweedy spinsterhood and insistence on her own brand of tea stamp her as eccentrically reliable in the best British tradition, is perhaps an even more striking example of what thirties 'Britishness' meant.

But already here there are implications of much that is repressed, or better, disavowed. The puritanism of British culture was a byword for a long time; the scandals caused by the *Lady*

British Disavowal and Repression

Chatterley trial, or the hints of nudity in some of the 'kitchen-sink' films, or Millicent Martin telling Roy Kinnear on *TW3* that his fly was undone – all a matter of twenty years ago! – seem inconceivably distant now, but – more to the point – they could not have occurred in any other European country except a Fascist dictatorship. We have seen how direct political intervention was practised (as it still is) in France, and how blacklisting and fear of such pressure-groups as the Legion of Decency acted as a powerful mechanism of self-censorship upon the American industry. The difference is that in France the sexual was largely subsumed under the political (for example, the banning of *La Religieuse*), and in America the industry found a multiplicity of ways round the limitations on sexual expression (for instance, the wisecracks of Mae West). There is no sense in which *The Lady Vanishes* gives the feeling of being a sexually repressed film; this is why the use of 'disavowed' seems to me more appropriate. To ascribe any kind of sexuality whatever to Miss Froy, at any time in her life, is almost literally unthinkable. The 'nun' who mounts guard over her on the train betrays herself through wearing high heels; this is not very 'English', and out of context would probably make us think of Buñuel rather than Hitchcock, but as Raymond Durgnat points out nuns are 'un-English' anyway, so that that particular perversity justifies itself outside the English context.[2] Charters and Caldicott are forced to share a single bed at the hotel,* and turn out to be sharing the only pair of pyjamas as well, so that each in turn has to shield the other's nudity as best he may. The ostensible humour derives from a generalized embarrassment and sense of put-upon dignity, but the way in which one of the men shields the other's nipples from the maidservant when she enters the room suggests more directly sexual implications even as the whole 'British' style laughingly disavows them.

If there is a characteristic British style of popular humour, it is one which rests upon this bizarre attitude towards sexuality – not quite constrained enough, to outward appearances, to make

---

*The notion of bachelor or 'unattached' males sharing rooms, if not sometimes beds as well, is an oddly persistent one in British cinema and television comedy; Tony Hancock was apparently worried that viewers would conclude from his sharing a bedroom with Sid James that their relationship was a homosexual one . . . No such qualms are reported to affect Morecambe and Wise, who like Charters and Caldicott palm the humour of *their* bed-sharing off onto general good-humoured embarrassment.

one think of it as 'repressed', but operating repression through its giggling half-avowals. The joke quoted at the beginning of the chapter is exemplary; the conjugation of mother-in-law, mammaries, and female underwear is clearly thought to be so screamingly funny in itself that the appallingly feeble pun will pass unnoticed. Each of the three items has a precise value. Mothers-in-law are the punishment men must endure for being foolish enough to marry (thereby admitting their sexuality publicly), and their blood-curdling ugliness suggests the souring and drying-up of sexuality – thus, by implication, what the hapless male's wife will look like in a few years' time. A glance at a Donald McGill seaside postcard will show that the British comic breast is constructed quite differently from its American counterpart (as in *The Girl Can't Help It*). American breasts are forward-thrusting objects of desire, British ones as often as not saggy; even when they are not they often contrive to deflect desire into embarrassment (think of how often on seaside postcards a middle-aged man leering at a young girl's breasts is chastised by his flabby termagant wife, with consequent red faces all round). The cardinal rule for female underwear is 'the baggier the better', and its contents are viewed exclusively from – literally – a rear elevation. It would be possible to construct a semantic table along the lines of the 'U/non-U' or 'In/Out' oppositions beloved of gossip-writers, as follows:

IN	OUT
Knickers	Pants
Botties	Arses
Boobs	Tits
Fannies (meaning 'backsides')	Fannies (meaning 'vaginas')

The items in the right-hand column are all too close to the overtly sexual to be admissible in the seaside-postcard joke or any of its derivatives. So tenacious and multiple are these that we shall see that even so self-consciously refined a film as *Kind Hearts and Coronets* bears their trace from time to time. The British have a reputation in French culture for sexual eccentricity bordering on the perverse (thus, sodomy and flagellation are both referred to as *le vice anglais*, and the debauchees in Georges Bataille's *Histoire de l'Oeil* enlist the aid of an English nobleman for their final sacrilegious outrage). It is easy to see

British Disavowal and Repression

how the whole complex of knicker-jokes and Charters and Caldicott in bed with each other (though emphatically not 'together'), of mothers-in-law and corporal punishment in schools, of Donald McGill and Lady Chatterley, could fascinate a very different culture by their bizarre otherness, and how the very mechanisms of disavowal that are at work here can be regarded as sexually perverse in themselves.

The associations that cluster round Oscar Wilde illustrate this well. The exchange with Whistler quoted at the beginning of the chapter (and popularized by a 'Monty Python' television sketch later reproduced on a record) derives much of its force from the reversal of Wilde's reputation for devastating originality. But it would hardly be accurate to say that it is by his wit alone that he is a household word to the British public, even many who have never read a line or seen a play of his. Wilde's reputation rests also on the 'devastating originality' (by conventional standards) of his love-affair with Lord Alfred Douglas; the perverseness of the wit and the 'perversion' of the sexuality have become inextricably intermingled in the popular Wildean myth. The phrase 'You will, Oscar, you will' is thus likely to evoke the inevitability of future sexual indulgence, despite the absence of reference to homosexuality in either the original context or the Python sketch.* 'Disavowal' is probably too strong a word here; it is more a matter of hints and connotations, a broad sideways wink at an audience who know what 'Oscar' is bound to get up to sooner or later. The opprobrium with which Wilde was treated after the court case, and the associations that have accumulated around him since, bespeak a culture not merely repressive, but at once morbidly and flippantly fascinated with that which it cannot, or dare not, face *seriously*.†

British film culture abounds in examples of the attitude I have just exposed. Michael Powell, until 1960 regarded as a major director, in that year delivered to his career a *coup de grâce* barely less lethal than that Carl Boehm/Mark Lewis delivers to his victims, by directing *Peeping Tom*. Mark, exposed as a child to various horrific experiences by his father, a psychologist

---

*It is interesting to note in this context that Graham Chapman, a leading member of the Python team, subsequently 'came out' about his own homosexuality in what was widely taken as proselytizing fashion.
†'Seriously' here is not to be equated with 'humourlessly'; Wilde is himself among the most 'serious' of British writers.

specializing in the study of fear (and played by Powell himself), takes in later life to inveigling female victims into his 16-mm film studio, there to 'shoot' them with a camera equipped with knife-blade and mirror, so that their death-agonies are not merely filmed, but reflected for them to watch. There has been a good deal written on the particular intensity with which the film distils the implicitly sadistic pleasure that is part of any cinematic experience. It is, however, the film's reception – at the time of its release and subsequently – that chiefly concerns us here. Ian Christie is quite right in pointing out that the storm of abusive criticism with which it was greeted cannot simply be derisively dismissed, reflecting as it does 'the *unacceptability* of the film when it appeared'.[3] But the manner in which this 'unacceptability' is re-articulated and denounced speaks volumes. Derek Hill in *Tribune* shifts genital fear to anal repugnance by saying that 'the only really satisfactory way to dispose of *Peeping Tom* would be to shovel it up and flush it swiftly down the nearest sewer'; C. A. Lejeune in the *Observer* refuses to name the actors in 'this beastly picture' (as Christie points out, thus 'censoring' the film); Campbell Dixon in the *Daily Telegraph* identifies a concern, 'fashionably' (?), with voyeurism; and David Robinson in the *Monthly Film Bulletin* accredits himself culturally by denigrating British cinema in favour of French literature, when he deplores the continued unavailability of the books of the Marquis de Sade along with the fact that 'it is possible, within the commercial industry, to produce films like *Peeping Tom*'.

These reviews provide rich terrain in themselves for a reading of sexual disavowal in British culture. The film's concern with the sadistic-voyeuristic elements in any film-watching – hence, by extension, with the unhealthiness of the artistic – is tackled only obliquely, by one or two critics; elsewhere, the kinds of diversion, censorship, denigration by association with the modish (to this day a favourite *Daily Telegraph* tactic), and cultural assertiveness exemplified above hold unchallenged sway. Nor can this merely be put down to the 'Philistinism' of British popular newspapers. In the first place, the opinions quoted come from largely 'respectable' (not to say intellectual) publications; secondly, there was an important force at work in the specifically intellectual life of Britain at the time by whose criteria *Peeping Tom* would certainly have stood condemned.

British Disavowal and Repression

This was the work that went on in the English Faculty at Cambridge (and thence elsewhere) around the journal *Scrutiny* and the teaching and writing of F. R. Leavis. That which was 'on the side of life' earned their approval; that which was seen as not so being was condemned; and this had nothing whatever to do with the explicit (or otherwise) depiction of sexuality. D. H. Lawrence is a far more sexually explicit novelist than Flaubert, but it is Lawrence's being 'on the side of life' as against Flaubert's supposed desiccated concern with art for art's sake that determines the verdicts pronounced on their work. This odd combination of vitalism and a reworked puritanism was arguably the dominant intellectual current in the Britain of the late fifties and early sixties. Here is no place to pronounce on the good and the harm (both unquestioned) it did. What is important is that a work of art which called the basis of art into question – which ventured to hint that between art and sexual psychopathology there might be connections deeper than generally acknowledged – was likely to get little quarter from either intellectual or popular audiences in the Britain of its time.

What is even more important is that, twenty years later and in a radically changed intellectual and social climate, *Peeping Tom* still has the status of a *film maudit*. More than any other single institution, the BBC bears witness to the change in British social attitudes in this period – from the pathologically uncontroversial 'Auntie' of 1960 to the alleged subversives' nest of 1980. Yet when the BBC presented an extensive Powell retrospective in the latter part of 1980 they happened to do so under the banner of Powell and Emeric Pressburger (his long-time collaborator). This may have been nothing but an innocent desire to pay homage to one of the best-known teams in British film-making; but the fact remains that it conveniently permitted, indeed obliged, the BBC to exclude *Peeping Tom* (alone among Powell's major post-war work), in the making of which Pressburger had had no part. The National Film Theatre, on the other hand, who organized a complete Powell retrospective two years before the BBC, showed all extant movies, *Peeping Tom* included. The conclusion – tempting if not quite inescapable – is that what was 'safe' for a metropolitan intellectual club was not so for the television screen. Not a great deal of imagination is required to see that this is indeed true (the monotonous howls of indignation from Whitehouse and her cohorts that on this occasion we were spared, by accident or design, almost induce a feeling of relief that the BBC programmed their season as 'Powell and Pressburger'). Twenty years after its original release – and twenty years after the *Lady Chatterley* trial, which banished the Puritan spectre from the realm of the written word – *Peeping Tom* remains taboo, for reasons linked with its sexuality at least as much as with its violence.

For a further example of sexual disavowal in British film culture, it is instructive to look at the manner in which the work of the theoretical journal *Screen* has frequently been received in Britain. *Screen* is published by the Society for Education in Film and Television, which receives financial aid ('public money') from the British Film Institute. This makes it a fairly obvious target for critical surveillance, to which no publication has *prima facie* reason to object. But the vituperativeness of much critical comment on SEFT in general and *Screen* in particular is practically unequalled in British film culture since the *Peeping Tom* furore. Why should this be so?

Screen operates (or perhaps it would be more accurate to say operated from the early seventies through to 1978 or thereabouts, when both its theoretical positions and its field of attention began to change) within a complex network of concepts drawn from modern, largely French, rereadings of Marx and Freud in conjunction with (or 'across') each other.* This has exposed it to some fairly predictable jibes on grounds of 'obscurantism' and 'pretentiousness' – neither allegation always entirely unjustified, but both drawing from the pragmatic distrust of theory so deeply embedded in British culture a force and credibility they could have found nowhere else. The British left in particular has consistently displayed a quasi-phobic aversion to theoretical concepts quite unparalleled in any other European left-wing culture (witness the fact that one of *Screen*'s most vitriolic critics is John Coleman of the *New Statesman*, the leading British left-wing intellectual weekly).

But times are changing, and this long-established distrust of itself is hardly a sufficient explanation for the venomousness of much criticism of *Screen*. I am tempted to suggest that another factor may well lie in precisely the tendency to disavow the sexual that I have been discussing. The work of such French

*For a fairly comprehensive introduction to this area of concepts, see Ros Coward and John Ellis, *Language and Materialism*.

writers as Jacques Lacan and Christian Metz, both very influential upon *Screen*, emphasizes the complex psycho-sexual charge of the act of *looking*, and hence the sexual (particularly voyeuristic) connotation of sitting in the cinema. Upon this foundation, which I have clearly had grossly to over-simplify, a theoretical discourse was constructed which deployed notions of castration, the Oedipus complex, fetishism, and the calling into question of stable sexed identity to account for the various forms of cinematic pleasure and to read a number of films in detail. Prototypical is the *Cahiers du Cinéma* reading of John Ford's *Young Mr Lincoln*[4] but the works of Godard, Oshima, Hitchcock, and, of course, Michael Powell have also been scrutinized, either in the pages of *Screen* or elsewhere by writers associated with it.

Screen thus mounted a double challenge, in the two areas where disavowal is most prevalent in British cultural life – theory and sexuality. The term 'disavowal', indeed, derives from Freud (the German is *Verleugnung*), and figures fairly prominently in *Screen*'s vocabulary. It would be perhaps unfair, but certainly possible, to fling back at *Screen*'s adversaries an *argumentum ad hominem* whereby the very frenzy of their denunciations would serve as proof of the validity of the magazine's positions; after all, does not every amateur psychologist know that we most fiercely criticize that which we suspect ourselves of possessing? Certainly it is interesting to find those associated with *Screen* and areas where its influence has been marked (notably the BFI Education Department) commonly referred to as 'pod people'. One main implication of this expression, drawn from the world of science-fiction, is that those to whom it refers reproduce asexually. I would not wish to suggest that whoever first used the expression in its SEFT context actually believed this to be true of the staff of either *Screen* or the BFI Education Department. However, it is curious to find, beneath other connotations such as lack of warmth and spontaneity, Dalek-like adherence to a preconstructed vocabulary and mode of speaking, and so forth, an aura of asexuality ascribed precisely to those who have been most active in exploring and developing the sexual aspects of pleasure in the cinema. Just as Derek Hill's phallic unease at *Peeping Tom* swivelled itself through a semicircle to evince (and evict) itself as anal revulsion, so one can find in the automaton-like qualities imputed to the 'pod people' evidence of an anxiety aroused in their critics which can be neither acknowledged nor simply brushed aside, and therefore emerges as projection (itself a form of disavowal).

I had better make it quite clear that I do *not* endorse every word that appears in the pages of *Screen*, and could indeed have devoted a good deal of space to criticism (albeit 'fraternal') of it; also, that I have no wish to impute deliberate personal malice to critics unknown to me personally, any more than I am capable of feeling, or entitled to feel, it towards them. My point is to be understood at a more general cultural level. The reasons why British culture (at least until very recently) has taken almost instinctive fright at any theoretical discourse, and continues to disavow sexual ambiguities it can neither abolish nor savour, are probably far beyond my individual scope, and certainly beyond that of this brief chapter. That these things are so, and that their conjunction in the pages of *Screen* has been met with criticism too vitriolic to be legitimately described as fair, is, I hope, clear. So too, I hope, is the fact that between the voluminous posteriors of Donald McGill and the 'pod-people' rhetoric of *Screen*'s adversaries, as between the 'mother-in-law's knickers' syndrome of British comedy and the excoriation of *Peeping Tom*, there is a connecting thread. This, among other things, we shall now find running through *Kind Hearts and Coronets*.

Kind Hearts and Coronets

The film relates to the rise of Dennis Price/Louis Mazzini to the dukedom of Chalfont, via the extermination of six of the eight people (all played by Alec Guinness) who stand between him and the title. Class ambiguity is at the base of the film. Louis' ruthless determination to attain the title springs from the continued snubbing of his mother (a Chalfont who has eloped with an Italian singer) by his ennobled relations, so that he can be seen as avenging not just one broken-hearted woman, but by extension all the wretched of the earth. Factors outside the film again colour readings of it, for if we know that Alec Guinness was knighted ten years after it was made, and that Dennis Price died a bankrupt in the Channel Islands, this seems to serve as a

British Disavowal and Repression

sadly piquant confirmation of the class position of the two men (or to be strictly accurate nine men and one woman) at the outset of the action.

Louis tells us near the beginning that whereas the British peerage traditionally descends through the male side only, the dukedom of Chalfont has been empowered to do so through both sides, 'for services rendered to His Majesty after his restoration by the Duchess'.[5] This is the first of a string of sexual allusions in the film whose witty urbanity can operate as an upper-class variant of disavowal. It would be perfectly possible for a first-time audience to miss the slyly cultivated allusion to Charles II's notoriously lecherous habits (he is in this respect an almost mythical figure as in every sense the antipodes of puritanism and restraint), and thus also to the fact that Louis' 'illegitimate' claim to the title is in fact legitimated by 'illegitimate' sexual commerce. It is widely current to speak of the look (or play of looks) in cinema; for *Kind Hearts and Coronets* it seems more apposite to speak of the film's *wink*, directed at the audience paradoxically not from the visual image but from the spoken commentary. The disjunction between the two is, as Charles Barr points out,[6] a key feature of the film; from our point of view it acts to reinforce the sense of disavowal, for all too often something perfectly innocuous will be going on before our eyes while the subversive or sexual charge, as here, will be concentrated on the soundtrack. There is not an image in the film, divorced from its spoken context, that is in any conventional sense (sexually at least) 'shocking'; the informed listener is far more exposed to this aspect of the film than the casual spectator.

The first contemporary appearance of sexuality occurs in the schoolroom, when Louis tells us *à propos* the Commandments that 'as to the Seventh Commandment, I was hardly of an age to concern myself with it, although I was old enough to be in love'.[7] This introduces us to Joan Greenwood/Sibella, Louis' first (and for all we know last) love, later described as 'pretty enough in her suburban way',[8] but throughout the film endowed with an erotic charge that gives the lie to such a dismissive statement. Louis entertains hopes of winning her hand through his meteoric rise in the draper's where he is constrained to work, culminating in 'ladies' underwear at 35/-'.[9] This joke works on two levels: first as a stylish up-market version of knickers humour (in no other culture could ladies' underwear have functioned so unquestionedly as a cue for laughter): and secondly, via the hope – implied, but not in so many words expressed – that this latest promotion will act as Louis' passport into feminine underwear of a somewhat less impersonal kind. Initially, this hope is rebuffed, for Sibella has decided that she will marry upper-class albeit slightly shady Lionel, and after rejecting Louis' proposal tells him that she is going to bed. He replies: 'You will'[10] (a textual wink in Oscar's direction?), and so, in due course, with both himself and her pledged bridegroom, she does. The beginning of the film has shown us Louis, in prison under sentence of death, writing his memoirs as tenth Duke of Chalfont, so that we know from the start that his apparently insane ambition will be realized, and the whole film is thenceforth likely to be read as an exercise in the realization of the improbable. So it is that the conjunction of the underwear promotion and the 'promissory note' Louis has the effrontery to write himself on Sibella's behalf works to reassure us that he will one day become her lover – all without so much as a hint of palpable impropriety.

His first murder is articulated round a classic of British popular humour, the honeymoon joke – here spiced by the strictly illicit nature of the 'honeymoon' between Ascoyne d'Ascoyne and a paramour never given any name more specific than 'girl in punt'.

There is a whole Oxbridge tradition (the d'Ascoynes, we have been told, always go to Trinity College, Cambridge) of punting on sunny afternoons, often as a means of winning over ladies on whom one might have designs. And then as now, Maidenhead (appropriately named?), scene of the *coup de grâce*, was something of an upper-class playground, though at the turn of the century it was not so much the locale for show-business personalities' riverside haciendas as a rather more up-market alternative to Brighton for the dirty (or at least illicit) weekend.

A whole variety of British innuendo has thus been deployed around Ascoyne d'Ascoyne's Thames side dalliance before its regrettable outcome. This is watched (by the audience aligned with Louis) at a distance; as the punt bearing the naughty couple disappears over the edge of a weir, Louis says: 'I was sorry about the girl, but found some relief in the reflection that she had presumably, during the weekend, already undergone a fate worse than death.'[11] At the risk of appearing ponderous, I shall venture a few observations on what is rightly one of the film's most cherished *bons mots*. The 'fate worse than death' referred to is clearly connected with disagreeable sexual experience (the phrase tends to evoke pagan times of pillage and rapine, when the menfolk of the invaded communities would be barbarously put to death while the women suffer a fate deemed to be even worse), and – at its limits at any rate – the raping of a virgin. While there is no suggestion that the nameless female was hitherto a virgin, and even less that what she undergoes at Ascoyne's hands is rape in any statutory sense, Louis' remark winkingly implies a degree of sexual pressure (not to say brutalization) on the part of the aristocrat, and its surface sexist condescension towards 'the girl' can be seen as masking an avenging of the more overt and odious sexism of the aristocracy already amply demonstrated in their treatment of his mother.

Charles Barr speaks of the Oedipal element in *Kind Hearts and Coronets*, exemplified in Louis' eradication of eight monstrously recurring 'father-figures', and in his own father also being played by Dennis Price (Louis thus being shown as 'married' to his own mother).[12] I would not wish to press this too far in the specific context of this study; but it is part of the the make-up of a 'gentleman' in any culture (and the expression, English in origin, has passed in varying forms into a number of European languages) not to tolerate denigration of his mother. Louis thus shows himself more of a 'gentleman' in the weir episode, among others, than those who may have the title but distressingly lack the kindliness or the style.

At the same time, his refusal to kiss the engaged Sibella piques her into saying: 'I like you when you behave like a cad'[13] – an expression initially coined at Eton and Oxford as a pejorative name for the townsmen, and on Sibella's lips evocative both of Louis' 'lower-class' position and of a certain sexual assertiveness. Only with Sibella is Louis exempt from the need to behave like an officer and a gentleman; thus, it is in the scenes with her that his sexuality can emerge most clearly. When he casually remarks to Lionel on the wedding morning: 'You're a lucky man, Lionel, take my word for it',[14] after spending the night with the bride, we are reminded of the observation attributed to an Oxbridge Head of House on the eve of the marriage of two of his ex-students: 'It'll be an excellent marriage, I know – I've slept with both of them.' The difference in the film, of course, is that, both for reasons of plot (it is vital that Lionel should suspect nothing) and because of the complex taboos on explicit avowal, the second part of the Head of House's statement is omitted, left to be supplied by an audience that may indeed unconsciously be reminded of it by the delivery and context.

It is after Sibella's marriage, and Louis' elimination of the photographer Henry d'Ascoyne, that the film's sexual intrigue becomes more complex. Louis decides to woo Henry's widow Edith (played by Valerie Hobson), to revenge himself on Sibella maybe, but also to prove his worthiness to become a member of the aristocracy. The tissue of innuendo around Sibella from now on becomes more complex, suggesting more the less it says. She visits Louis, on her return from honeymoon, in his newly acquired apartments in St James's (London's 'clubland', and traditionally an area inhabited by well-off bachelors), and to his inquiry about whether she enjoyed her honeymoon she replies. ' "Not at all." – "Not at all?" – "Not at all." '[15] The pouting delivery of Joan Greenwood makes understood by implication a great deal about the qualitites of Louis and Lionel as lovers. Later on, Sibella is eager to know what Mrs d'Ascoyne is like – a question which can be fitted into the same context as the honeymoon exchange – and, when Edith has accepted his proposal of marriage, Louis is led to reflect *à propos* Sibella that: 'Women have a disconcerting ability to make scenes out of nothing, and to prove themselves injured when they themselves are at fault.'[16]

The aphoristic distance of the generalization – Sibella lumped together with other 'women' and thus divested of her specific charge – begins to indicate a sexual and emotional distancing from her that is the result of self-interested scheming at least as much as of weariness. At the wedding reception, Louis has earlier reflected that 'revenge is a dish which people of taste prefer to eat cold,'[17] and the drop in temperature engendered by his observation on Sibella hints that mealtime may not be too far away. His libidinal energy, needless to say, has to be severely restrained with the d'Ascoynes; in order to expedite his murderous designs on the suffragette Lady Agatha, he finds himself carrying a suffragette banner in a demonstration, which in the context carries a certain charge of emasculation. Lady Agatha, in her admittedly extremely brief appearances, evokes two English stereotypes – the tweedy spinster *à la* Dame May Whitty,

British Disavowal and Repression

but also an area of disavowal at once crude and complex, the male in female clothing. From the pantomime Dame through to Danny LaRue, this is another staple of the British sense of humour which to this day can trigger off quasi-Pavlovian spasms of laughter from television-comedy audiences, in a manner unparalleled in any other culture. Lady Agatha is, among other things, *Kind Hearts and Coronets*'s passing tribute to the power of British drag.

Louis is able to take a revenge in many ways more satisfying than his path to the dukedom when Lionel comes desperately to appeal to him for financial help. Louis' cool refusal, Lionel's venomous outburst against Louis' mother and origins, Louis' striking him, and Lionel's subsequent suicide (misinterpreted as murder, for which Louis will later find himself in the dock), complete the revenge begun on the eve of Lionel's wedding to Sibella. Lionel takes his first love: Louis cuckolds him. Lionel insults his mother: Louis effectively drives him to suicide. Libidinal and emotional energy takes ever more vengeful paths, sexuality always cloaking itself in a mantle of gentlemanly style.

This reaches its apotheosis in Louis' murder of the major obstacle between him and the dukedom – the old Duke Ethelred, of all his victims the only one who can truly be held personally responsible for the affronts to Louis' mother. Ethelred is lured (shades of *La Règle du Jeu*, and love-tangles and class-hatreds of a very different kind?) into a mantrap he himself has set for catching poachers. He is then, quite calmly, eliminated with a shotgun – a killing far more coldly ritualized than the others, and preceded by a speech outlining Louis' reasons for revenge and the way in which he has gone about it. What comes to the fore here (and we may think both, in a very different context, of Bataille's English nobleman and of the success the film enjoyed in France under the inspired title of *Noblesse Oblige*) is *sadism*, absent or jocularly evicted from the five previous murders but making its return here in chilling fashion. The Duke's greater age, his position as a trapped animal (and poacher of that which should rightfully not have been his), and above all the measured tones in which Louis pronounces the sentence of death upon him, all combine to give the effect of a quasi-judicial execution of which de Sade himself need not have been ashamed. The 'British style' has the paradoxical result of at once masking and, by implication (especially for a non-British audience), bringing out the sadistic elements in Louis' enterprise.

At the end, Louis, under sentence of death, has nevertheless a possibility of escape, held out by the untiringly blackmailing Sibella. She will obligingly 'produce' her late husband's 'suicide note' if Louis undertakes upon his release from prison to do for Edith what he has already done for six d'Ascoynes and make her, Sibella, his Duchess in her place. Louis has earlier said of her: 'We serve each other right, you and I',[18] and the two of them now, in their competing unscrupulousness and perversity, are reminiscent of Choderlos de Laclos's ultimately homicidal libertines, Valmont and Madame de Merteuil, in *Les Liaisons Dangereuses*. Louis accedes; he leaves the prison to be greeted by both Sibella and Edith (at least one of whom is presumably soon to get a disagreeable surprise); and then comes the famous concluding episode, where a journalist from *Tit-Bits* stops him outside the prison gate to ask for publication rights to his memoirs:

'(puzzled) My memoirs?
(seeing the point) My memoirs!
(as a horrified thought strikes him) My memoirs!
(and again) My memoirs!'[19]

This ranks among the most celebrated ambiguous conclusions in cinema, though the version shown in America included, at censorial behest, a final shot showing Louis' manuscript (left behind in his cell) being handed over to the authorities. How we read the ending as originally filmed is a difficult question to answer. Plausibility would presumably dictate that Louis' attempt to turn back to his cell would alert the governor's suspicions and lead to his being found out; but, whatever criteria we adopt in reading the film, plausibility is unlikely to rank very high. Barr puts it well: 'Is there any question which is more likely or which Hamer intends us to think is more likely?'[20] The flashback structure of the film has assured us from the beginning that Louis' ambition will be fulfilled – it is almost a case of: 'I wish I could murder my way to the dukedom' – 'You will, Louis, you will.' The thrust of the film is towards the idea that crime does pay; but, even supposing Louis to go free in the moments after the film's final shot, the homicidal (and sexually rooted) intrigue of *Kind Hearts and Coronets* could not be said

British Disavowal and Repression

to end there. Of the three main characters who gather outside the prison at the end, one at least will be marked down for violent death, whether it be Louis (through Sibella's denunciation if he fails to eliminate Edith), Edith, or Sibella (depending on which way Louis' exterminatory wind happens to blow). It is a testimony to the perverse elegance and power of this most remarkable of film comedies that it closes, with urbane good humour and a final ambiguous wink, upon a classically tragic situation – the prospectively homicidal triangle of jealousy.

Kind Hearts and Coronets, then, disavows to the last – as in some degree do all films, but in a manner specifically British in its elision of, or oblique winking references to, the libidinal and the sexual. The fleeting appearances of knicker-jokes and drag, the thematic play made with the differences (sexual and other) between aristocrat and parvenu, the Wildean qualities to which so many commentators on the text have drawn attention – it should not be difficult to see how these fit into the complex of British humorous attitudes towards sexuality. They do rather more than that, of course; the success of the film in France, as compared with the failure of McGill postcards or mother-in-law jokes to make their mark on that culture, is proof enough of that. But that very success again fits into the Gallic view of British upper-class perversity already mentioned, one that British culture itself has never been able to indulge because of the mechanisms of disavowal so deeply rooted within it.

It remains now to situate *Kind Hearts and Coronets* within the context of the time of its making – something I have carefully refrained from doing so far the better to bring out its more general properties. The film was denounced by Lindsay Anderson, who became a leading proponent of the more 'realist' Free Cinema, as 'emotionally quite frozen'; cryogenic immersion might be a more tenable process to have suggested than outright freezing, for as we have seen the film's superficially glacial elegance acts to disavow its complex perverse exploration of libidinal energy. There were also contemporary reasons for this, for as Barr points out the Ealing ethos (carrying over from that of wartime Britain) was predicated on the acceptance of 'restraint on sex drive and ambition and class resentment'[21] – all restrictions *Kind Hearts* sweeps aside with sublime urbanity. In this context, the film appears as a 'holiday' from the 1940s British spirit of self-denial, and perhaps its 'cryogenic' quality and the mechanisms of disavowal it operates were necessary to stave off the threat of shocked or horrified reaction. That no suggestion of the *Peeping Tom* outcry greeted *Kind Hearts* cannot simply be put down to the fact that one is a 'comedy' and the other a 'psychological horror film'. Beneath the genre labels, it is two different kinds of spectator involvement and positioning that are at work – Powell's film forcing us into compliance with a horrifying equivalent of our own position in the cinema, Hamer's appealing (albeit ghoulishly) to the traditional view of the 'old' British aristocracy as comic grotesques* and maintaining a distance that begins to shock only when it winks in the direction of sadism. The difference between avowal and disavowal, measured in the very different reception given to the two films, tells us a great deal about how British culture will tolerate its sexuality being constructed, on celluloid and elsewhere. Powell had more integrity than to have wished that he had never made *Peeping Tom*: had he done so, the only reply would have been: 'You did, Michael, you did', and a reassurance that the fault lay in an audience and a whole culture, not in one director.

---

*As in Vivien Stanshall's recent *Sir Henry at Rawlinson End*, to say nothing of the Goon Show's Major Bloodnok – an old-school aristocrat *manqué* if ever there were.

Part Four

Inscrutability and Reproduction – Four Major Japanese Directors

The text does not 'comment' on the pictures. The pictures do not illustrate the text.[1]

Nothing has better served detractors of 'Japanese culture', nor more keenly embarrassed the bulk of its Western admirers, than the ease, indeed the eagerness, with which the Japanese, at certain periods of their history, have adopted techniques and concepts of foreign origin . . . The ease with which she has adopted Western techniques in modern times is merely additional proof that her only 'aptitude' is a mimetic one.[2]

The irreducible 'otherness' traditionally ascribed to oriental cultures in the West is an ambiguous phenomenon, and nowhere more so than in any consideration of Japan. This is at least partly because Japan's status as an 'oriental' country is itself somewhat ambiguous. Noël Burch has pointed out[3] that, though geographically undoubtedly oriental, Japan is unique not merely in the Far East, but in the whole non-Western world, in never having been colonially subjugated, and that unlike all other Asian nations she never experienced what is known as 'oriental despotism' – an authoritarian and centralized tier of hierarchies that found its economic determination in the need to mobilize large numbers of labourers to irrigate vast tracts of land. Many of the features of Japanese society most familiar in the West today (often, precisely, through their reproduction in the cinema) likewise go to reinforce a perception of the country as 'inscrutably' oriental, yet at the same time radically different from other oriental cultures – that is to say, radically *other* by occidental and oriental norms alike.

Between the dispossessed knights of *The Seven Samurai*, forced by changing social conditions into the role of mercenaries, and the bounty-hunters of the American Western, there is a certain amount in common, which is why John Sturges was able to 'adapt' Kurosawa's film to a Western setting as *The Magnificent Seven*. Yet the success of both films with occidental audiences is attributable as much to the piquancy of the cultural distance between them as to their similarities of plot and action. Kurosawa's later films, *Yojimbo*, and *Sanjuro*, made after *The Magnificent Seven*, illustrate this point well. They deploy elements traditionally associated with the American Western (such as the small township tyrannized by rival gangs in *Yojimbo*), in a manner that Western audiences might savour almost as high camp. At least, this is the conclusion one is tempted to draw from the Italian 'spaghetti Western' remake of *Yojimbo*, *A Fistful of Dollars*, in which the elements of parody and stylization come to dominate the film. But there is a whole other dimension to the Kurosawa films of which most Western audiences are likely to be unaware – that of the social changes that took place in Japan in the 1860s, the transition from an insular, proto-feudal society to one based on mercantile and cultural expansion, which accounts for the desperately precarious existence of the samurai (the displaced feudal class) and the tyrannical power the merchants are able to wield. To be aware of this is not to lose the iconographic comedy of the film, doubtless important to Kurosawa if only because it widened his prospective audience, but rather to begin to see how the humorous elements and those, closely associated, of visual spectacle are related to a context whose otherness and unfamiliarity for us would be second nature for a Japanese audience.

Another reason for the tendency in the West to foreground the comic elements in Kurosawa's samurai films is the widespread reputation of the Japanese for copying. The quality of Japanese cameras, hi-fi equipment, motor-cars, and watches is never, of course, called into question; but their presence in the homes, in the garages, and on the wrists of those who some years ago (if only as a result of post-war prejudice) would have bought

German, British, or Swiss is still a source of cultural embarrassment. Partly it is, of course, a result of shifts in international alliances, which have transformed the Japanese from bloodthirsty military adversaries to partners in capitalist trade. But the sense of grievance at the 'pirating' of European or American originality persists, and it is possible to see in the amused reception of the Kurosawa films in the West a comic deflection of this, a recognition that, supposedly not content with large-scale plagiarism of things Western, 'they' – the Japanese – have actually plagiarized the 'Western' itself, and that the West is mature and broad-minded enough to return the compliment by replagiarizing the Kurosawa films in its turn.

But this is dependent upon a conception of plagiarism so culturally specific that we need to make a considerable effort to remind ourselves that its validity in Japanese culture is at best questionable. The other side of the alleged Japanese facility in copying Western inventions is their 'sense of tradition',[4] an expression that certainly has meaning in any study of Japanese culture but requires historical and philosophical elucidation. Notions of creativity and authorship in the West tend to suppose that scientific 'discovery' and artistic 'invention' stem from *autonomous* individual creativity. I emphasize the word 'autonomous' to make it clear that it is not the notion of individual creativity that I am seeking to deny, but the quasi-miraculous aura with which it is surrounded in the West. Creators in Japan, whatever else they may be, are not gods; they operate within a system that has tended to deny any individual or proprietorial rights over the work produced. Historically the roots of this attitude may be traced back to the long period of feudal isolation from the outside (or at least the Western) world, making possible the continuation of ritual observances far beyond what in a more open society would have been considered their allotted span. Economically, the renowned Japanese skill at copying and improvement found its justification in the need for revival after the Second World War, and may well also have drawn upon the absence of an individual artisanal tradition of the kind prevalent in the West. The French sociologist Jean Baudrillard, in *Le Système des Objets*, has established a helpful distinction between what he terms 'models' – individual, 'customized' articles produced for a select few – and 'series' – articles that are mass-produced and widely available – and emphasized that this distinction is a constantly changing one, since yesterday's costly 'model' is today's inexpensive 'series', and those who wish to assert their élite status are therefore compelled to go seeking after new 'models' to maintain their distance from the masses. A whole economy of consumption has thus been constituted around the ethic of individual creativity, and it becomes easier to see how Japanese culture, historically less dependent on the conception of originality, could leave the tedium of selecting the 'models' to the West and concentrate on the more lucrative task of transforming them rapidly into 'series'.

The radically diffferent conception of individual identity in Japanese culture is also linked with the important role of Buddhism, particularly Zen Buddhism, within it. During almost 700 years of rule by the *shogun* caste of warrior aristocrats, the main intellectual and political focus for the lower classes was provided by the various Buddhist congregations and monasteries, and it was in these that the complex stylization of *noh* theatre, with its masked actors and spoken commentary on the action,* began to develop. Unity rather than separateness, the unsayable rather than the theologically definable, the interconnectedness yet uniqueness and irreplaceability of even the most (by Western standards) trivial and repetitious experience – these are important strands in Buddhist thought, and it should be plain enough how alien to them the Western conception of authorship and proprietary rights over one's 'creation' is. If in the predominantly Christian West the creator is God, and the book (or indeed film) a form of Holy Writ, such an analogy is very difficult to sustain in a society whose most influential religious philosophy has neither God nor scriptures.

That Zen, among whose primary principles is an extreme respect for all organic life, coeisted for so long with a feudal society whose violence escapes the label of barbarous only (if at all) through its extreme codification – that a pacific organicism and a stratified cruelty should have inhabited, as their descendants continue to inhabit, the same culture – is an extraordinary paradox, perhaps the one which most fully gives the measure of how 'Japan'† is other than both 'Eastern' received norms of

*This had a profound influence on the work of Eisenstein, the first major theorist of cinematic stylization and derealization.
†I use the inverted commas to emphasize that I am referring to a constructed image of a society and a culture rather than to some preexistent geographical reality.

Four Major Japanese Directors

passivity and tranquillity and 'Western' received norms of virile, self-assertive aggression. While the Buddhist influence, apparent in if not actually ventilated through the films of Ozu and Mizoguchi, clearly operates in the former direction, the 'Japan' of the war years, whose barbarism reached such mythical heights that only the ultimate purgation of the nuclear bomb was deemed sufficient to eradicate it, and that of the post-war years, with its guards forcibly cramming bodies onto the Tokyo underground and its phallocentric culture drunken and shallowly unfaithful beyond the wildest dreams of any Hollywood macho, have equally clearly worked to produce an image of a Japan more ruthlessly occidental than the occident.

Where the two stereotypes not so much meet as overlap is in the ritual violence so much part of any image of Japanese culture. The ritual disembowelling with one's own sword – *seppuku*, known in the West as 'hara-kiri' – characteristic of disgraced knights in the feudal period, and spectacularly revived by the writer Mishima Yukio in 1970, is the most widely known example of this. However, the torments inflicted in Japanese prison-camps in the Second World War and the internecine violence and torturing of the ultra-left guerrilla Red Army in the early 1970s also go to reinforce the image of the Japanese nation as given to horrific, yet contained, violence in the name of this or that ideal. It is the combination of the savagery of the violence itself and the willing acquiescence of the victims (who may, as in *seppuku*, themselves be the inflictors) that confronts Western sensibilities with a dizzying paradox.

Both the point I have made about ritual violence and that about the unimportance of individual creativity clearly link into Japan's past as feudal-warrior nation and into her present as the world's supposed master plagiarist. In terms of the types of cinematic style her culture evolved, they are also connected with a lack of stress on *interiority* quite unlike anything in the culture of the West. The privileged status of the close-up in much Western cinema has to do with the notion that a human being's interiority – what he or she is 'really like' – is most immediately and reliably available through the face. The revelation of Kane's secret to the audience at the end of *Citizen Kane*, the progressive destruction of our and Joan Fontaine's illusions about what kind of person Daphne du Maurier's and Hitchcock's *Rebecca* was, the disclosure in *All About Eve* that it is Anne Baxter/Eve and not Bette Davis/Margo who is the truly ruthless character – these serve as examples of how the linear movement of narrative in dominant Western cinema can fulfil a role analogous on a wider scale to that of the close-up, by stripping away misconceptions about a character and thus placing the audience in the position of psychological voyeurs. The stress Zen places on continuity and flow rather than on separateness in depth, the manner in which the structuring of so much Japanese social activity around the ritualized observances of the tea-ceremony or the bar appears (for Western spectators at any rate) to blur or deny the specific individuality of those involved, even the classic Western difficulty in reading a Japanese countenance (a sophisticated variant of 'they all look the same to me') – these factors combine to give the concept of interiority a much reduced importance in Japanese cinema. Close-ups, in the films of Ozu, are as likely to be of objects as of countenances; the large-scale battle-scenes in Kurosawa's samurai films, the use Mizoguchi makes of long- or crane-shots in the kind of charged situation which would lead Western film-makers to think more readily of employing close-ups, the stylized tableaux of much of Oshima's work – these are other cinematic strategies that bespeak a culture whose notions of individuality, identity, and interiority are in important ways different from ours.

It may, in the light of what has just been said, appear peculiarly perverse to announce now my intention of organizing this chapter around four *auteur*-directors. The reasons for this are threefold. In the first instance, while the director in Japan was, unsurprisingly, slow to emerge as the acknowledged *auteur* of his work, his power and influence once he had done so grew apace. Joseph L.Anderson and Donald Richie wrote in 1959 that '. . . the Japanese film director, as far as having the final say goes, is among the strongest in the world',[5] and subsequent developments appear to have done little to jeopardize his position. This may appear surprising, but the tenacity of hierarchies of power in all walks of Japanese society no doubt has a great deal to do with it. Secondly, the 'friction' or paradox produced by examining a cinema predicated on very different notions of individuality through one of the more sophisticated Western individualist conceptions of the creative process is perhaps a useful tactic for bringing out how Japan appears to the Western audience through its cinema. And thirdly, it is far truer of the

Japanese cinema than of any of the others treated in this book that it is accessible in the West only through the output of a very limited pantheon of *auteurs*. Of the four whose work I have elected to discuss – Ozu Yasujiro, Kurosawa Akira, Mizoguchi Kenji, and Oshima Nagisa – only Kurosawa's *oeuvre* was at all widely known in Britain at least before the early seventies. Ozu and Mizoguchi are represented for the British audience by mere tips of their authorial icebergs (even this has been true for Mizoguchi – with the periodic exception of *Ugetsu Monogatari* – only since the mid-seventies), and the *cause célèbre* of *Ai No Corrida* in 1975 was required for Oshima's name to become known outside a very narrow circle. I shall write about what is available and perceptible, but the vast number of lost or irrecoverably 'buried' films should constantly be borne in mind while reading my analysis.

Ozu Yasujiro

The number of 'Ozus' available for consumption by Western audiences is manifold. When *Tokyo Story* first made his name known in the West, it was the supposed quietistic strain in his work, the calmly loving attention to details of everyday life and the virtual disappearance of 'plot' in the occidental sense, that was emphasized, by such writers as Penelope Houston.[6] On this reading, Ozu thus emerged as a kind of super-realist, whose fascination with what other directors would have considered minutiae provided his audiences with a privileged vantage-point on a different culture.

But other methods of reading his work have since become available. It is virtually unique among that of major directors in demonstrating no interest whatever in romantic or passionate love. If Ozu's characters marry, one is sometimes tempted to feel, it is only in order that they can produce family-units for him to anatomize. Psychoanalytic and feminist writing, the work of directors such as Godard and Oshima, and developments in the writing of history that attach great importance to areas not habitually thought of as 'historical' (as in the work of Michel Foucault), have in recent years spread the awareness that the family is a nexus of politics as well as of sexual-romantic love.

Ozu manages the considerable feat of devoting probably more footage to the structures and rituals of family life than any other major director while effectively ignoring the two areas – the sexual-romantic and the political -- that converge upon it. For Anderson and Richie, in *The Japanese Film*, this is the source of much of his interest; they see him as in this respect 'the Japanese's Japanese' *par excellence*, somehow 'above politics', and so concerned with the iconography of the visual image that he remained for a long time resistant to sound. For writers concerned with the artist as human being, the exclusion of romantic love from Ozu's films fascinatingly complements his own lifelong bachelorhood and intense relationship with his mother.* For those interested in what a film implies, or suggestively omits, as much as what it says, Ozu's 'miniaturism' is significant because of the way in which the wider political and historical life of Japan percolates into his films through fissures and crises in the family structure (an example is the son's call-up in one of Ozu's only two wartime films, *There was a Father*).

For Paul Schrader, on the other hand, interested not in the social but in the transcendental, that cinema which attempts to reach what he describes as 'the Wholly Other, [which] once perceived, cannot be limited by culture'[7] finds in Ozu one of its foremost exponents. Here it is the movement towards transcendental calm which is seen as fundamental to Ozu's work, and Ozu the apparent quietist thus reappears in this reading, but in a very different light from that of Penelope Houston. His preoccupation with stasis rather than motion, the tendency of his camera to dwell on spaces just vacated, or not yet occupied, by the characters, and the fact that his 'plots' are barely perceptible as such now appear as indices of that which lies ineffably beyond the fabric of Japanese social life, rather than as documentation of its details. The ritualistic form of that which is filmed (a life bounded by four locations, the restaurant, the home, the office, and the bar) interacts with the hieratic quality of the camera's position and movement to point constantly beyond the banal realities represented to a resolution of the conflict between the old and the new (as in the 'generation gap' of *Tokyo Story*) via a stance of ironic acceptance. This approaches what is known in

---

*An interesting parallel here is with the life and work of another, very different, supreme master on what by conventional Western standards is a highly limited canvas, the Argentinian writer Jorge Luis Borges.

Four Major Japanese Directors

Zen as *mono no aware*, an attitude described by W. H. Auden as the state where 'we laugh because we simultaneously protest and accept'.[8]

The social and the metaphysical find a meeting-point here, so that the transcendental conservatism implicit in the resolution of Ozu's conflicts has often been seen as placing him on the far-right wing of Japanese cultural politics. This is an over-simplification; his anatomy of the families in *Tokyo Story* or *The Brothers and Sisters of the Toda Family* is no less stringently critical for not pointing to any 'revolutionary' destruction of the institution, and a film such as *Ohayo/Good Morning* is eloquently condemnatory of the influence of Westernized commercialization on Japanese life. Ozu's own political views appear chiefly conspicuous by their absence, though his wartime experiences were apparently unpleasant and there is no suggestion of militarism in his work. 'Above politics', the label proposed by Anderson and Richie,[9] is perhaps misleading; it might be more profitable to speak of Ozu as 'beside' politics, addressing himself to problems that fall within the domain of the political but without using a political (or even in any real sense a religious) discourse.

The two Ozu films I shall now briefly analyse have been chosen partly because they are both (particularly *Tokyo Story*) sporadically viewable in the West and partly because both well illustrate the coexistence of the poignant and the playful in his work. *Tokyo Story* deals with an elderly couple's visit from the south of Japan to their married children in Tokyo. The children, a doctor and a beautician, are too caught up in their own concerns to devote much time to their parents, and the couple's supposed 'treat', an excursion to a spa resort, is clearly a tactic for getting them off their children's hands. Tomi, the mother, is taken ill in Osaka on the way home, and dies; Shukishi, the father, is shown in an archetypal Ozu situation/shot (the two are so closely linked that it is almost impossible to speak of them separately), alone but for his loneliness in the house of his daughter-in-law, who is a war-widow. Central to an understanding of the film is the importance of filial piety – of 'honouring thy father and thy mother' – in Japanese culture. One of its cruellest ironies for a Japanese audience is that only the daughter-in-law displays this in her genuine concern for the old couple.

At the end of the film the daughter-in-law confesses to Shukishi that she finds life without a husband extremely difficult, and he counsels her to marry again. Loneliness is not the only common denominator here. In a society where even nowadays thirty per cent of marriages are arranged, and where one keystone of the social edifice is the gracefully subordinate position of women, an unmarried female (whether 'bachelor' or widow) is by definition considered of inferior status, at least temporarily in one respect a failure. For widowers such as Shukishi, an approved life-style does exist, movingly presented in Ozu's last film, *An Autumn Afternoon*. It consists of visits to and from surviving members of the family (an option which obviously becomes considerably less attractive for Shukishi after the trip to Tokyo), and nostalgic *sake*-drinking sessions with similarly widowed friends. There has been a foretaste of this when Shukishi gets embarrassingly drunk on his last night in Tokyo; there is little joy or conviviality in his intoxication, which is both cause and effect of his growing sense of awkwardness with his own children. Thus neither Shukishi nor Noriko (the widowed daughter-in-law) really has anything much to look forward to, at least unless Noriko accepts the advice to remarry or Shukishi learns to love falling off bar-stools in the company of his peer-group. The immutable hierarchy of Japanese family and social life, whatever Ozu's own attitude towards it, has left these characters very little choice. The end of the film is often adduced as a key cinematic example of *mono no aware*, but how far this is the film's and how far that of the characters is a very difficult question to answer.

Tokyo Story, like almost all Ozu's films, is shot from what is known as the *tatami* position. *Tatami* means 'rush matting', and a noteworthy feature of Japanese interiors is that chairs are much less common than in the West; people always used to kneel, crouch or even lie (especially after much *sake*) on rush mats at floor-level.* Much attention was drawn to the low position of the camera when Ozu's films first appeared in the West, and exaggerated stories circulated about how having to lie on cold floors to film caused several of his cameramen to leave his service for the sake of their health. In fact the *tatami* position used to be as 'natural' for a Japanese scene or audience as is for

*This is less and less true in modern Japan, where the *tatami* position has indeed served to date Ozu's work.

Four Major Japanese Directors

us the 'sitting' position from which most Western scenes of social intercourse are shot. Its strangeness for us is an index both of the otherness of Japanese social life, even to the very position in which it takes place, and of the exceptionally high proportion of interior and family scenes in Ozu's films.

The role of drinking in Japanese life is also well brought out, particularly via the widowed characters and office-workers, in Ozu's work. It is quite normal in Japan for white-collar workers to go straight to a bar from the office and reel home at any time up to midnight, not just now and again but several nights a week, without feeling any sense of guilt or obligation towards their wives (or towards the tube employees who will have to mop up the copious pools of vomit that begin to appear on station platforms from mid-evening onwards). This indicates both how phallocratic a society Japan is and how the tenacious male ritual of drinking has taken on a grotesquely exaggerated form with the pressures of competitive industrialization. Home drinking is also important, as a kind of sacrament that binds the males together on the *tatami*. Ozu's own consumption of *sake*, when discussing ideas with his scriptwriters, was impressive, and this is mirrored in his films. But, as has already been pointed out, Shukishi's drunkenness is a sorrowful parody of the male-group ritual so important in Japanese life and culture. The widower in *An Autumn Afternoon,* on the other hand, has learnt – if only *faute de mieux* – to enjoy his regular sessions in the bar. Alcohol for Ozu is a mournful as well as a communicative sacrament.

Nothing so far in these comments on *Tokyo Story* has suggested the presence of a celebratory or playful element in what is undoubtedly for the most part a film of elegiac resignation. But this emerges strikingly in one sequence, when the old couple take their first tram ride through Tokyo. Here the camera positions us near (though not *with*) them inside the tram, and their excitement and joy in the bustling movement constitute the film's only moment of exhilaration; one could even in a Joycean sense speak of a small-scale 'epiphany', the positive recognition of and fusing with a moment of existence that is perhaps first cousin to – or the obverse of – *mono no aware*. It is so easy to stress the resigned stasis and elegiac quality in this best-known of Ozu's films that the 'life' imparted by the tram sequence is often neglected. It is important, apart from anything else, as a subtle suggestion that the modernization of Japan in general and Tokyo in particular may not have had entirely negative results.

This potential ambivalence emerges in altogether more light-hearted fashion in *Ohayo/Good Morning*, different from *Tokyo Story* in being a comedy and a colour film. It is essentially a remake of an early Ozu silent, *I Was Born, But . . .*, made twenty-seven years earlier – a fact which has caused Noël Burch to dismiss the film rather scathingly in a footnote as indicative of the 'sheer exhaustion of the later years'.[10] But the later film says a good deal about the invasion of traditional society by consumerism. The two small boys round whom the film centres demand a television set from their father; when he refuses to buy them one, they go on 'silence strike', taking his irritated injunction to shut up so literally that they will not even utter the 'good morning' of the film's title to their neighbours, thereby almost provoking a major disagreement between families. This appears at first to be little but a rather silly running gag more suitable to a children's television series; but the importance of ritual politeness in Japanese society, and the implied clash between this and the desire for Western machines of mass entertainment of which the television is the classic example, is in fact quite subtly brought out. The ending evokes 'happily ever after' rather than the *mono no aware* of *Tokyo Story*; but, just as the sorrowful resignation of the earlier film had found some counterweight in the exhilaration of the tram-ride (a positive memory for Shukishi to put alongside that of Noriko's kindness), so the resolution of *Ohayo*, with the television proudly installed and the 'magic words' spoken once more, hints at a rather melancholy regret that the urge for consumer goods and mass entertainment could have won a battle in which the old observances were used as weapons.

Ohayo is one of Ozu's most apparently 'realist' films, with its absence of static passages, its use of colour, and its close attention to the day-to-day life of the street. But its most notorious, and seemingly most 'realistic', feature should serve to warn us against too ready an acceptance of this manner of reading the film. The greatest joy in the lives of the two small boys, and of the friends with whom they daily walk to school, is to fuel themselves with pumice-powder so as to fart as abundantly as possible. Twice in the film this leads to disaster, when the same boy fouls his underpants. Their father also farts frequently, thereby earning his sons' admiration (women are even less likely to do so in Japanese than in occidental culture). But what we actually hear on the soundtrack is unlike any fart ever perpetrated by human bowel; it resembles much more the squeaking noise produced by blowing through a small leaf folded in half. This may simply have been because more 'realistic' farts (examples of which occur in Mel Brooks's *Blazing Saddles* and Marco Ferreri's *Le Grande Bouffe*) might have been considered shocking. But it also seems to me to go along with the general evasion, if not denial, of 'realistic' bodily functions most evident in Ozu's lack of interest in sexuality. We know that the noise we are hearing is supposed to be a fart (even if we have not previously read about the film) because of the characters' attitudes towards it and the 'accident' that befalls one boy in the first fart sequence. But it is in fact a stylized representation – as stylized in its own way as anything in the work of this most stylized of directors. I would not wish to carry this too far and see in the underpants that fly flag-like from the clothes-line in the film's final image a banalization of *mono no aware*, or a transcendental meditation on the unreliability of the human sphincter; but it does serve to show how in Ozu's work the trivially smutty, the culturally critical, and the melancholy never far beneath the surface of things can converge even in the most superficial of moments. That Ozu's camera-style is determinedly non-conflictual (there is little sharp cutting back and forth or montage of opposites) does not mean that conflict is not everywhere present, and – perhaps – everywhere ironically reconciled, in his work.

Kurosawa Akira

Rashomon's award at the Venice Film Festival of 1951 has counted for more than any other single event in promoting consciousness of Japanese cinema in the west. The film is alleged to have been made along intentionally 'Western' lines in order to gain wider acceptance; if so, all that can be said is that Kurosawa's strategy was spectacularly successful, for the subsequent availability (restricted though it has been) of films by Ozu, Mizoguchi, and others would surely never have come about without the breakthrough of *Rashomon*. The supposed 'Westernness' of the film rests largely upon its use of a narrative device all too familiar from the early days of 'experimental' fiction – the recounting of one series of events successively from different subjective points of view. This has combined with the rather suspect uplift of the ending, where the woodcutter's taking care of an abandoned child is clearly meant to efface the sombre picture of human mendacity and corruption, to give *Rashomon* a somewhat dated quality of liberal relativism – the Japanese Bloomsbury Group film? But such a reading is likely to be predicated upon ignorance of specifically Japanese references that make the film far less 'Western' than is popularly supposed.

The plot concerns the death of a samurai in a wood, in a dispute over his wife with a bandit played by Mifune Toshiro (later to become *the* Japanese actor for Western audiences). Each of the four participants – the woodcutter, the wife, the bandit, and the samurai – gives their version of the story; the samurai is able to do this from beyond the grave because he speaks through a medium. This is presented as being quite natural, which given the importance of ghosts in Japanese narrative art, and the tendency to portray them naturalistically rather than as wafts of ectoplasmic transparency, would have made sense to a Japanese audience. Similarly, the husband's *seppuku* or hara-kiri, at least in his own posthumous version of events, is a response to his wife's arousal by the bandit, whom she begs to kill her husband and run away with her. It would have behoved a samurai thus sexually shamed to commit suicide in this way. The classic Western phallocratic response – murder of the wife, her lover, or both – would simply not have been culturally appropriate.

This shows how inevitable it is that a Western audience will select the aspects of a Japanese film that it can most easily adopt and 'naturalize', passing by others whose impression on a Japanese audience would be just as great. Clearly only complete cultural 'bilinguality' can fully obviate this problem, endemic in some degree to any foreign product (including for a London audience an American film or one from the North of England); but it recurs more often with some films than with others. The 'Japaneseness' of Ozu's work is so pronounced that those viewers who do not find it unwatchable are likely to foreground it in their readings of the films, though the 'realist' reading outlined

Four Major Japanese Directors

above comes close to an assimilation of Ozu to Western cinematic norms. Kurosawa's work, on the other hand, at least those examples of it most widely available in the West, lends itself much more readily to 'Westernized' readings. For *Rashomon* these even went so far as to claim, quite inaccurately, that the film had not gone down well in Japan, being too 'Western' to appeal to audiences there.

The director himself finds such assertions intensely irritating, saying that he has not 'read one review from abroad that hasn't read false meanings into my pictures'.[11] While this leaves open the question of what exactly constitutes a 'false meaning', it does warn us of the pitfalls attendant upon any attempt to treat *Rashomon purely* as an exercise in narrative relativity or *The Seven Samurai purely* as a stylish oriental plagiarization of the Western. The latter reading, common also for *Yojimbo* and *Sanjuro*, is favoured by the tendency, already noted, for the Western to be collapsed into a series of permutations on a finite number of basic plots; Will Wright's schemes would doubtless be applicable to the Kurosawa films, but this applicability would be only partial, for as well as being formal they are also socially and historically determined, and in the samurai movie we are dealing with a quite different social and historical situation. Such a warning may seem absurdly self-evident, but much critical writing about and audience reaction to Kurosawa's work shows how necessary it is in even this brief attempt to take stock of some of the ways in which Japanese society constructs itself through its cinema.

The action sequences in *The Seven Samurai* are, quite justifiably, what has primarily endeared the film to several generations of audiences, and the techniques Kurosawa adopted to shoot them (multiple cameras, huge quantities of footage ninety per cent of which was then discarded) certainly bear more relation to the big-scale Hollywood epic style of filming than to the tranquil economy of an Ozu or a Mizoguchi. But the film's tragi-comic tone has other roots than the combination of excitement, bravado, and slapstick in its action scenes. The samurai sub-genre of the *jidai-geki*, or period film, enjoyed widespread popularity for several reasons (apart from its success on the foreign market). Joan Mellen has suggested that at a time when the Japanese film market was controlled – as it still is – by large reactionary conglomerates, the *jidai-geki* could act as a 'surrogate for modern Japan and a subtle cultural device for simultaneously assailing modern Japan, with her current evils, while showing the ingrained historical, social, and structural roots of the problems at hand'.[12] This was of course a doubtful enterprise in many ways; just as those involved with the *Till Death Us Do Part* series on British television found that the benighted class and racial prejudices of Alf Garnett, which the series was intended to satirize, actually secured the programmes an enthusiastic contingent of Garnett-admiring viewers, so films such as Kurosawa's, which satirize Japan's feudal and militaristic past, inspired nostalgic reactions in many of their audiences.

The samurai of the film's title are in actual fact *ronin* – masterless knights left in a precarious position by the civil wars of the fifteenth and sixteenth centuries, which led to the disintegration of centralized rule, the rise of the infantry, and the splitting-up of Japan into a number of autonomous feudal dominions. The peasant village in which the action is set decides to hire the samurai to defend it against marauding bandits, a class-alliance that in any other circumstances would have been unthinkable. The samurai were agents of the system of feudal oppression, from which the peasants would have suffered hardly less than from the depredations of the bandits. The hesitation in the village before the samurai are approached is therefore understandable, and the apparently unfriendly attitude of the peasants towards the samurai at the end is not just a meditation on the constant ingratitude of humankind, but a reaction grounded in a very specific set of historical circumstances. The samurai are the peasants' traditional enemy, and their tragedy at the end of the film is that there is literally no place for them. As knights on horseback, they are servants without masters; and it is culturally unthinkable that they will lay down their arms and become peaceful tillers of the soil. The film thus has an elegiac quality that is profoundly ambiguous. Is Kurosawa (known for his progressive and anti-militarist opinions) pointing to the good, even lovable, qualities that his *ronin* show as evidence of what was repressed and disavowed in their samurai days? Or is he rather yielding to the temptation to humanize, and thus in some way to glorify, a particularly barbaric period of Japanese history?

The lack of historical perspective from which Western audiences are likely to suffer in their dealings with Japanese cinema emerges strikingly when we consider that it would be perfectly possible to watch and enjoy *The Seven Samurai* and *Yojimbo* without in any way taking the measure of the 250 or so years that separate the action of the two films. *Yojimbo* is set in 1860 – eight years before the restoration of Imperial rule and the designation of Tokyo as the capital. Japan had known something like four centuries of total economic and cultural isolation, under a rigorously stratified yet also decentralized feudal structure, a period characterized by chauvinistic nationalism, an ethic of revenge that to contemporary Westerners appears barbarous, and a complex network of duty and privilege. The imperatives of economic progress were what finally broke this archaic system down, with the opening of the ports to American vessels in the 1850s. The feuding silk-merchants in *Yojimbo* reduce the township to a state of terror by their ruthless tactics; in this respect the town is, to quote Joan Mellen, 'a microcosm of nascent capitalism',[13] where naked economic rivalry can and does thrive with no holds barred because there is no stable political system to govern its operation. The drafting of Mifune (again, *ronin* rather than samurai) into the service of the townsfolk takes on a new significance in this light. Many disinherited or disillusioned samurai were to be among the leaders in the movement towards centralization and stabilization that culminated in the Imperial restoration of 1868; a class whose roots and traditions were unequivocally feudal actually transplanted a large part of its allegiance to the installation of a bourgeois-capitalist régime, albeit one with substantial surviving feudal elements.

None of this is to denigrate the film's comic qualities, nor the manner in which it parodies the heroics of the 'straight' samurai picture (most notably in the person of one preternaturally enormous warrior who marches through the film clutching an equally enormous mallet, mercifully never used). Rather, it is to point out how a whole nexus of social, economic, and historical factors contributes to *The Seven Samurai* and to *Yojimbo*, and how an understanding of them contributes to our appreciation of two films more complex than is often realized.

Mizoguchi Kenji

Japan's society, as has already been mentioned, is among the most phallocratic in the world; yet the work of the two directors we shall now look at ranks, in very different ways, among the most cogent and moving cinematic protests against the mistreatment of women. This is remarkable enough in the work of Oshima, but even more so in that of Mizoguchi, who filmed over a thirty-year period during which the subordination of women went very largely unquestioned. Yet he was able to produce a number of remarkable films (only a very small number of which, alas, have ever been commercially shown in the West), whose consistently sensitive treatment of women and anger at their oppression emerges the more strikingly for being set in a variety of historical contexts. The action of the nine Mizoguchi films I have seen takes place across a period of almost 800 years, from the end of the twelfth century (*Sansho Dayu*) to the period immediately after the Second World War (*Portrait of Madame Yuki*). This made commercial sense, for the national past has always been good box-office in Japanese cinema; but it may also have helped Mizoguchi to avoid censorship and repression by cloaking his criticisms of Japanese society in historical garb, and it certainly provides a graphic illustration of how closely woven into the fabric of Japanese life barbaric treatment of women has always been. We have already seen how elements of feudal oppression survived well into the period of capitalist industrialization. Nowhere was this truer than in the treatment of women.

But it would be an oversimplification to speak of Mizoguchi as though he were simply an intelligent Japanese male with a conscience, which he assuaged by making films with nobly put-upon heroines. For the visual and narrative strategies he adopts are as important to his films as the 'stories' these tell; indeed, it is particularly difficult in Mizoguchi's work to disentangle the two.

The best way of illustrating this is to look at three important films: *The Story of the Last Chrysanthemums* (made in 1939, set in the late nineteenth century); *Portrait of Madame Yuki* (both made and set in 1930); and *Sansho Dayu* (made in 1954, set in the last years of the twelfth century). I have stressed, as before in a rather different context for much French cinema, the gap between the time of the films' making and that of their action to

Four Major Japanese Directors

emphasize how Mizoguchi deals with features common to all periods of Japanese society without losing sight of their historical specificity. His strategies of filming are an important element in this. With hardly any other film-maker are questions of camera-position and breadth or depth of visual field so closely linked with the view of a particular society we derive from his films.

Noël Burch illustrates this admirably in his study of *The Story of the Last Chrysanthemums*.[14] This he regards, on the basis of largely formalist criteria, as Mizoguchi's masterpiece, drawing particular attention to the unusual length of individual shots and to the important alternation between 'montage within the shot' and 'the "scroll" approach'. By the former, Burch (following Christian Metz) means the combination of juxtaposition of a number of different elements within one image;* by the latter, the 'unrolling' of the action through lateral tracking-shots, analogously to the pictorial scrolls characteristic of Oriental graphic art. What is the importance of this for the view of Japanese society in the film?

Eric Rhode has said of Mizoguchi's potters in *Ugetsu Monogatari* that 'it is unclear whether these men are at the centre or the periphery of the world'[15] — a judgement that at first appears the reverse of historical, but which in fact embodies an important historical ambiguity. The distance between the date of the action of *Ugetsu* and that of the making of the film is about 300 years. By a variety of cinematic strategies of which the use of long- and crane-shots is the most significant, this distance is at once brought home to us and symbolically abolished. Brought home, because of the disparity between the privileged 'reality' of what we see on celluloid and the elements of period stylization (as in the ghost story), and because camera and historical distance 'rhyme' with or echo each other; symbolically abolished, through the apparent lack of specific historical positioning to which Rhode alludes and the possibility of reading the camera's distance from the action – especially in the famous final crane-shot – as in some sense transcending the temporal and historical order (*mono no aware* of a different kind?)

The strategies referred to in *The Story of the Last Chrysanthemums* work in a similar way. When the actor Kikunosuke

---

*The use of 'deep-focus' eulogized by André Bazin in the work of Welles and Renoir is but one particular form this strategy can take.

first meets the nursemaid Otoki, who is to become his support and inspiration, they are presented to us in such a way that they never occupy more than the bottom third of the frame. A great deal of oriental graphic art is not centred upon individual human beings, omitting them altogether (as in the Western landscape) or, more subversively from an occidental point of view, suggesting their relative unimportance in the scheme of things through their relatively diminutive stature. Such a reading will only partially 'take' for a film so obviously anthropocentric as Mizoguchi's. What the mode of filming suggests is rather that these two characters whose destinies we are to follow for the next two hours are important, but not so important that they will be able to escape the pressures of forces more powerful than they – historical, social, and perhaps even in Paul Schrader's sense transcendental. As the camera dollies along with the characters, it passes street-sellers and houses – the physical and social landscape against which their story will be played out.

The final sequence of the film is also noteworthy in this respect. Kikunosuke has achieved fame and status as an *oyama*, or female impersonator (as in Shakespeare's England, women were not allowed to appear on the stage in Japan); Otoku, who has given him devoted support despite ostracism from those around him for her lowly birth, is seriously ill with tuberculosis. At the end, Kikunosuke is called way to the triumphal boat-procession by which the people of Tokyo used to pay homage to their favourite actors. The sequence cuts between Kikunosuke at the prow of the boat that is bearing him along victoriously and Otoku's classically 'Bohemian' death in a garret. The final shot shows us Kikunosuke looking elegiacally up as though aware of Otoku's death and the price at which his success has been bought.

Several points emerge from both the narrative and the filmic strategies here. First, that an imitation, surrogate femininity is privileged by the society of the film* over the 'real' femininity that has to die to nourish it. Secondly – a point also finely made in a very different context by *La Règle du Jeu* – that it is when class and gender prejudices and hostilities converge that the greatest harm is likely to be done, but also that the harm is most easily concealed – banished to the fastness of a garret, or

---

*I use this term with intentional ambiguity, to refer either to the society portrayed or to that in which the film was made.

papered over by the rancid residue of aristocratic style. Thirdly, that the camera's distance from the dying Otoku, contrasted with its homing in on Kikunosuke's 'triumph', is productively ambiguous. Is it a reflection of the fact that Kikunosuke, as man and actor, has all along received far more attention from society than Otoku? Or, as Burch suggests,[16] a partial reversal of the Western use of the close-up at moments of emotional intensity, a 'reverse codification' of the notion of interiority already discussed and hence a possible denial of it? Or (and it should be borne in mind that actors in *noh* and *kabuki* theatre often spoke from behind masks they held in front of their faces) a literal penetration to the 'man behind the mask' that is Kikunosuke, so that it is as though a 'Western' ethos of individual interiority and an 'Eastern' one of transcendental distance were alternating in the film's final moments? Whichever of these readings we adopt (and nothing prevents a combination of all of them), it should remain clear that all are rooted in an ambiguity of time closely connected with Rhode's ambiguity of place, and that whether we 'are' in 1885, 1939, or somewhere 'between' or 'above' them is centrally important to the film. Mizoguchi's apparent penetration to the heart of eternal verities can also be seen as an exploration of the ambivalence of history.

Portrait of Madame Yuki, an example of Mizoguchi's 'contemporary' film-making, is much more obviously relevant to modern Japan. The heroine's husband is a figure of such chauvinistic grossness as to be almost a caricature, though reports of contemporary Japanese male behaviour (as in the after-work drinking bouts) suggest that this is a rather genteel Western-metropolitan view; the intensive modernization of Tokyo in particular has given the always phallocratic bias of Japanese society a peculiarly nasty kind of updating. The husband's mistress is a cabaret singer; Yuki's admirer is too reticent to overcome scruples about her marriage-vows; and her suicide by drowning is denounced by her hitherto adoring maid in the final shot, where the camera moves down towards the waters of the lake as the maid calls her dead mistress a coward. The downward movement – the reverse of that which characterizes the endings of many Mizoguchi films, such as *Ugetsu* – may have been intended to mime the film's more immediate contemporary relevance. At any rate, such is certainly its effect; the historical remove produced as the camera soars above the fields at the end of *Ugetsu* is here nullified, any suspicion of *mono no aware* disappears, and the ripples on the lake are clearly tremors that will one day shake the smooth surface of Japanese society.

How little that surface had fundamentally changed since feudal times is shown in *Sansho Dayu*. The feudal code in operation in the twelfth century requires the wife of the insurgent lord to abandon her husband and take her children back to her own family – his legal and personal bonds have been annulled by his act of disobedience, and her views on the matter are simply of no consequence. Her son, Zushio, is the focal point of the film; his barbarism as custodian of the prison-camp, where he brands a seventy-year-old man with a red-hot iron and abandons an old woman to die of hunger, is brought to an end by a childhood memory, and he absconds to be recognized as an aristocrat and thus succeed his father. His quasi-Tolstoyan attempt to eliminate slavery from his own lands leads him to resign his governorship of the province, and in the film's final sequence he finds his mother, now old and blind, beside the sea. The camera distance here stresses perhaps even more laceratingly than at the end of *Ugetsu* the centrality of the characters, yet also their relative insignificance – for there must have been literally thousands of others like them in Japanese history, and the barbarities of the war years were sufficiently recent to act as a reminder that democratization and industrialization had not eradicated such tragedies from Japanese society. *Sansho Dayu* is both a classic period film and a work of great relevance to post-war Japan – both a transcendental meditation and a political and historical statement.

Oshima Nagisa

Oshima has drawn the subject-matter of most of his films from the history and present social reality of Japan. *The Boy*, about the traumatized confusion of a boy who acts as breadwinner for his entire family by feigning injury in road accidents and getting substantial cash compensation, was based on a contemporary newspaper story. *The Ceremony*, a work that anatomizes the rituals of Japanese family life more thoroughly than anything I know of outside Ozu, also functions as a complex allegory of the

Four Major Japanese Directors

political life of modern Japan, in which the characters and their bonds and feuds represent tendencies and events in Japanese history. *Death by Hanging* derives much of its force from the fact that the condemned central character is a Korean; the Japanese colonization and exploitation of Korea led to an attitude of contempt rooted in national guilt very similar to the anti-North-African racism in France or that directed against Indians and Pakistanis in Britain.

In all these films, the role of the family is central and its potential for oppression and exploitation is constantly emphasized. *Ai no Corrida/Empire of the Senses* extends these considerations into the realm of extra-marital sexuality. The central characters are a restaurateur and the geisha for whom he leaves his wife, and with whom he copulates unceasingly and inventively. The film's erotic frankness led to a harassment that turned out to be highly lucrative; it was shot in Japan but edited in France to circumvent censorship, and its effective banning in its country of origin has been more than offset by the vast amounts it has grossed at the box-office elsewhere (particularly in France, where it clearly profited from the tendency to condone the erotic and censor only the very explicitly political).

It also led to a concentration upon the clearly non-simulated quality of the sex scenes that detracted from a consideration of the film's highly political nature. To attempt to do more than summarize this here would be both presumptuous and superfluous, for Stephen Heath has presented this aspect of the film admirably in his article *The Question Oshima*. But it is still necessary to refer to it, for otherwise any consideration of how Japanese society appears in the film is likely to disappear beneath a blanket assertion of its otherness – the spontaneously demanding yet at the same time highly ritualized quality of the love-making, the sense of self-abandonment coexisting with self-awareness that reaches its apotheosis at the end, where she strangles him with his consent at the moment of orgasm. The film's final titles tell us that it happened in 1936 – a reminder that it is based on an actual episode, but also a complex allusion to a whole tissue of historical circumstances.

1936 was the year of an attempted military coup in which three ministers were assassinated; some of these had the previous evening been guests at the first showing of a sound film in Japan. The following year was to bring war with China – the beginning of the end for Japanese militaristic ambition. There is only one overt reference to this in the film (a parade of soldiers under the banner of the rising sun), yet indirectly it pervades the whole work. The climate of instability and violence in which such a love-affair was possible; the impact of Westernization (such as the talking pictures) upon traditional Japanese social norms; the crisis of domination, military and sexual, which the lovers' behaviour clearly acts out; the undermining of male dominance exemplified when the female takes more and more of the sexual initiative, the hieratic quality of the performances (in every sense of the word), their feeling of codification-unto-death – all these factors cohere around the year 1936, so that the final title is not just a reminder that what we have just seen had 'really' happened, but a statement about the entire context of the film and its possible relevance to an industrially revitalized, but still abominably sexist, Japan. The otherness of *Ai no Corrida* is specific and historical as well as culturally transcendent.

References

(All translations from the French are by the author.)

Introduction
1. *Sociologie du Cinéma*, p. 94
2. See *Signs and Meaning in the Cinema*, p. 97

Part One
1. *Capitalism and Freedom*, pp. 14/15
2. *Ibid.* p. 4
3. *Genre*, p. 19
4. *Screen Education*, Winter 1979/80, p. 42
5. Quoted in *Genre*, p. 51
6. *Underworld USA*, p. 18
7. *The Rise of the American Film*, p. 12
8. *Ibid.* p. 226
9. *We're in the Money*, p. 97
10. *Essais sur la Signification au Cinéma*, vol. II, p. 74
11. *Movie-Made America*, p. 188
12. *Popcorn Venus*, p. 135
13. *Toms, Coons, Mulattoes, Mammies, and Bucks*, p. 82
14. *An Introduction to American Movies*, pp. 55–7
15. *Westerns*, pp. 18–22
16. *Underworld USA*, p. 35
17. *A Biographical Dictionary of the Cinema*, p. 416
18. From a magazine in the Film Archive of the Museum of Modern Art, New York

19. *Hollywood in the Forties*, p. 28
20. *Journey down Sunset Boulevard*, p. 268
21. *The Citizen Kane Book*, pp. 374–5
22. *Ibid.* p. 438
23. Quoted in *Orson Welles*, p.42
24. *Focus on Howard Hawks*, p. 91
25. *Report on Blacklisting: I – The Movies*, p. 82
26. *Capitalism and Freedom*, p. 19
27. *The Musical*, p. 23
28. *The Strange Case of Alfred Hitchcock*, p. 329
29. *Humphrey Bogart*, p. 137
30. *The Strange Case of Alfred Hitchcock*, p. 301
31. *A Biographical Dictionary of the Cinema*, p. 522
32. *Don Siegel: American Cinema*, p. 37
33. *Ibid.* p. 74
34. *Sight and Sound*, Summer 1978, p. 192
35. *Sight and Sound*, Spring 1979, pp. 82–5

Part Two

1. *Le Scribe*, p. 127
2. *Recherches sur le Vocabulaire du Général de Gaulle*, p. 97
3. *La Société Française à travers le Cinéma, 1914/1945*, pp. 9–10
4. *Cinéma et Histoire*, pp. 11–12
5. For the factual information and quotations in this section I am indebted to *La Société Française à travers le Cinéma, 1914/1945*
6. *Ibid.* p. 78
7. *Jean Renoir: the World of his Films*, p. 57
8. *Cinéma et Société Moderne*, p. 26
9. *Un Maître du Cinéma: René Clair*, p. 150
10. *Les Malédictions du Cinéma Français*, p. 166
11. *Ibid.* p. 161
12. *Albert Camus, 1913/1960*, p. 102
13. *Les Malédictions du Cinéma Français*, pp. 14–15
14. *Jean Renoir: the World of his Films*, p.116
15. *Ibid.* p. 128
16. *Peasants into Frenchmen*, pp. 439–41
17. *The Rules of the Game*, p. 154
18. *Ibid.* p. 55
19. *Ibid.* p. 168
20. *Muriel*, p.102
21. *May 68 and Film Culture*, p. 93
22. *Cinéma et Histoire*, p. 56. (Jacques Soustelle was one of the leaders of the 'French Algeria' movement.)
23. *Ibid.* p. 57

References

Part Three

1. *A Critical History of British Cinema*, p. 133
2. *The Strange Case of Alfred Hitchcock*, p. 152
3. *Powell, Pressburger, and Others*, pp. 53–9, from which Press quotations are also taken
4. Reprinted in *Screen Reader –1*, pp. 113–152
5. *Masterworks of the British Cinema*, p. 202
6. *Ealing Studios*, pp. 123–4
7. *Masterworks of the British Cinema*, p. 203
8. *Ibid.* p. 207
9. *Ibid.* p. 209
10. *Ibid.* p. 213
11. *Ibid.* p. 213
12. *Ealing Studios*, pp. 126–7
13. *Masterworks of the British Cinema*. p.215
14. *Ibid.* p. 220
15. *Ibid.* p. 226
16. *Ibid.* p. 237
17. *Ibid.* p. 220
18. *Ibid.* p. 249
19. *Ibid.* p. 264
20. *Ealing Studios*, p. 130
21. *Ibid.* p. 120

Part Four

1. *L'Empire des Signes*, p. 5
2. *To the Distant Observer*, p. 31
3. *Ibid.* pp. 29–30
4. *Ibid.* p. 30
5. *The Japanese Film*, p. 346
6. *The Contemporary Cinema*, pp. 146–7
7. *Transcendental Style*, p. 55
8. *Ozu*, pp. 7–8
9. *The Japanese Film*, p. 363
10. *To the Distant Observer*, p.277
11. *The Japanese Film*, p. 376
12. *The Waves at Genji's Door*, p. 85
13. *Ibid.*p. 23
14. *To the Distant Observer*, pp 230–36
15. *A History of the Cinema*, p. 389
16. *To the Distant Observer*, p. 236

Glossary

This gives a brief selection of technical terms that figure in the text, or are likely to crop up in serious modern writing about the cinema, along with brief working definitions.

auteurism   the theory that ascribes prime creative responsibility for a film to its director

castration complex   the fear, attributed by Freud to all males in Western society, of being castrated by the father (or a father-figure) as punishment for incestuous designs on the mother

crane-shot   one taken from a great height, often moving upwards away from the scene being filmed; a famous example is the end of *Ugetsu Monogatari*

deep-focus   the camera plunging deep into the illusory 'third dimension', of the screen, endeavouring to give an impression of recessed depth and permitting several different events to take place within the one frame (examples: *The Best Years of our Lives*, *Citizen Kane*)

180°   the convention, rarely transgressed at any rate in Western narrative cinema, that the camera should not cut through an angle greater than 180° from one shot to the next lest it jeopardize the spectator's sense of spatial position

phallus   a term whose use in the Lacanian development of Freudian psychoanalysis is as complex as it is current. It relates not to the biological penis, but to the father-figure (endowed both with a penis of his own and with the supposed power to amputate that of transgressors) as source of Law and legitimacy, an unchanging and unchallengeable ground for knowledge and certainty. First made widely current by *Cahiers du Cinéma* in their reading of *Young Mr Lincoln*

Glossary

point-of-view shot   one taken from the point of view of a particular character, so that what s/he sees is also what the audience sees

soft-focus   the blurring effect often used in 'romantic' cinema to focus attention on one part of the frame, or indicate e.g. drunkenness or sudden falling in love on the part of a character

superstructure   the 'scaffolding' of politics, philosophy, ideology (a much-disputed term in this context), and culture in any society

tracking-shot   one where the camera moves along, either past static objects or past a landscape itself in motion. A particularly notorious example is the traffic-jam shot near the beginning of *Weekend*

zoom   where the camera homes in on a particular person or object, thereby bringing it into the centre of attention. The final glimpse of the 'Rosebud' sledge in *Citizen Kane* is a kind of zoom

Bibliography

This includes all non-fiction books referred to in the body of the text, listed alphabetically with author and publisher details.

Albert Camus, 1913/1960, Philip Thody, Hamish Hamilton 1964
Big Bad Wolves, Joan Mellen, Elm Tree Books 1977
Biographical Dictionary of the Cinema, A, David Thomson, Secker & Warburg 1975
Capitalism and Freedom, Milton Friedman (with Rose Friedman), Chicago University Press 1962
Cinéma et Histoire, Marc Ferro, Denoël/Gontier 1977
Cinéma et Société Moderne, Annie Goldmann, Denoël/Gontier 1974
Citizen Kane Book, The, ed. Pauline Kael, Paladin 1974
Contemporary Cinema, The, Penelope Houston, Penguin 1963
Critical History of British Cinema, A, Roy Armes, Secker & Warburg 1978
Donald Siegel – American Cinema, Alan Lovell, British Film Institute 1975
Ealing Studios, Charles Barr, Cameron & Tayleur/David & Charles 1977
Écrits, Jacques Lacan, Éditions du Seuil 1966/1971
Eighteenth Brumaire of Louis Bonaparte, The, Karl Marx, Progress Publishers 1954
Empire des Signes, L', Roland Barthes, Flammarion 1970
Essais sur la Signification au Cinéma, Christian Metz, Klinksieck 1968/1972
Film Sense, The, Sergei Eisenstein, Faber & Faber 1943
Focus on Howard Hawks, ed. Joseph McBride, Englewood Cliffs/Prentice Hall 1972
Genre, Steve Neale, British Film Institute 1980

Bibliography

Godard: Images, Sounds, Politics, Colin MacCabe (with Laura Mulvey and Mick Eaton), British Film Institute 1980
Histoire de l'Oeil, Georges Bataille, Pauvert 1967
Historical Novel, The, Georg Lukàcs, Peregrine 1969
Hollywood Babylon, Kenneth Anger, Straight Arrow books 1975
Hollywood in the Forties, Charles Higham/Joel Greenberg, Tantivy 1968
How to Read Donald Duck, Armand Mattelart, New York 1975
Humphrey Bogart, Alan G. Barbour, Pyramid Publishers 1973
Images of Alcoholism, Jim Cook/Mike Lewington, British Film Institute/Alcohol Education Council 1980
Introduction to American Movies, An, Steven C. Earley, New American Library 1978
Japanese Film, The, Joseph L. Anderson/Donald Richie, Grove Press 1960
Jean Renoir: The World of his Films, Leo Braudy, Doubleday 1972
Journey down Sunset Boulevard, Neil Sinyard/Adrian Turner, BCW Publications 1979
Language and Materialism, Ros Coward/John Ellis, Routledge Kegan Paul 1977
Maître du Cinéma: René Clair, Un, Jean Charensol, Table Ronde 1952
Malédictions du Cinéma Français, Les, Francis Courtade, Alain Moreau 1978
Masterworks of the British Cinema, ed. John Russell Taylor, Lorrimer 1974
May 68 and Film Culture, Sylvia Harvey, British Film Institute 1978
Movie-Made America, Robert Sklar, Random House 1978
Muriel, Jean Cayrol, Éditions du Seuil 1963
Musical, The, ed. Richard Dyer, British Film Institute Educational Advisory Services 1975
Orson Welles, Joseph McBride, Secker & Warburg 1972
Ozu, Donald Richie, University of California Press 1974
Panorama du Film Noir Américain, Raymond Borde/Étienne Chaumeton, Minuit 1955
Peasants into Frenchmen, Eugen Weber, Chatto & Windus 1977
Popcorn Venus, Marjorie Rosen, Coward/McCann/Geoghegan 1973
Powell, Pressburger, and Others, ed. Ian Christie, British Film Institute 1978
Qu'est-ce que le Cinéma?, André Bazin, Éditions du Cerf 1978
Recherches sur le Vocabulaire du Général de Gaulle, Jean-Marie Cotteret/René Moreau, Armand Colin 1969
Report on Blacklisting: I – The Movies, John Cogley, Fund for the Republic 1956

Rise of the American Film, The, Lewis Jacobs, Harcourt Brace 1939
Rules of the Game, The, Jean Renoir, Lorrimer 1970
Screen Reader –1, Society for Education in Film and Television 1977
Scribe, Le, Régis Debray, Grasset 1980
Signs and Meaning in the Cinema, Peter Wollen, Secker & Warburg 1972
Société Française à travers le Cinéma, 1914/1945, La, René Prédal, Armand Colin 1972
Sociologie du Cinéma, Pierre Sorlin, Aubier Montaigne 1977
Strange Case of Alfred Hitchcock, The, Raymond Durgnat, Faber & Faber 1974
Studies in European Realism, Georg Lukàcs, Merlin Press 1972
Système des Objets, Le, Jean Baudrillard, Gallimard 1968
Toms, Coons, Mulattoes, Mammies, and Bucks, Donald Bogle, Viking Press 1973
To the Distant Observer, Noël Burch, Scolar Press 1979
Transcendental Style in Film, Paul Schrader, University of California Press 1972
Underworld USA, Colin McArthur, Secker & Warburg 1972
Waves at Genji's Door, The, Joan Mellen, Pantheon Books 1976
We're in the Money, Andrew Bergman, New York University Press 1971
Westerns, Philip French, Secker & Warburg 1977
Work of Dorothy Arzner: Towards a Feminist Cinema, The, ed. Claire Johnson, British Film Institute 1975

Filmography

This lists, in alphabetical order, every film referred to in the text (with the exception of a very few never shown in Britain, for which I was unable to get details from the BFI Information Department). It gives the film's title, the name of its director(s), its date, and a selection of its leading actors/actresses arranged in alphabetical order, where this appears appropriate. Dates are taken from Cawkwell and Smith's *World Encyclopedia of Film*, except for films not listed therein or released after the book was published, for which my source has usually been the BFI Information Department. *Films on Offer*, published annually by the BFI, provides a list of films which are available for hire on 16-mm (the size used by college and film-society projectors). I did not think it worthwhile to include this information here, since it would have been out of date by the time the book appeared.

A Bout de Souffle, Jean-Luc Godard, 1959: Jean-Paul Belmondo, Jean Seberg
Across the Pacific, John Huston, 1942: Mary Astor, Humphrey Bogart
Adam's Rib, George Cukor, 1949: Katharine Hepburn, Spencer Tracy
Afrique 50, René Vautier, 1955
Age d'Or, L', Luis Buñuel, 1930: Gaston Modot
Ai No Corrida/Empire of the Senses, Oshima Nagisa, 1975: Fuki Tatsuya, Matsuda Eiko
Air Force, Howard Hawks, 1943: John Garfield
Alamo, The, John Wayne, 1960: Laurence Harvey, John Wayne, Richard Widmark
All About Eve, Joseph L. Mankiewicz, 1950: Anne Baxter, Bette Davis, Gary Merrill, George Sanders

All That Heaven Allows, Douglas Sirk, 1955: Rock Hudson, Jane Wyman
Alphaville, Jean-Luc Godard, 1965: Eddie Constantine, Anna Karina
Amants, Les, Louis Malle, 1958: Jeanne Moreau
American Graffiti, George Lucas, 1973: Candy Clark, Richard Dreyfuss
Angels with Dirty Faces, Michael Curtiz, 1938: James Cagney, Dead End Kids, Pat O'Brien
Annie Hall, Woody Allen, 1977: Woody Allen, Diane Keaton
A Nous la Liberté, René Clair, 1932: Raymond Cordy, Henri Marchand
Apocalypse Now, Francis Coppola, 1979: Marlon Brando, Dennis Hopper, Martin Sheen
Arroseur Arrosé, L', Louis and Auguste Lumière, 1895
Atalante, L', Jean Vigo, 1934: Jean Dasté, Dita Parlo, Michel Simon
Autumn Afternoon, An, Ozu Yasujiro, 1962: Ryu Chishu
Bande à Part, Jean-Luc Godard, 1964: Anna Karina
Best Years of Our Lives, The, William Wyler, 1946; Dana Andrews, Fredric March, Harold Russell
Biches, Les, Claude Chabrol, 1968: Jean-Louis Trintignant
Big Parade, The, King Vidor, 1927: John Gilbert
Big Sleep, The, Howard Hawks, 1946: Lauren Bacall, Humphrey Bogart, Martha Vickers
Birth of a Nation, The, D.W. Griffith, 1915: Lillian Gish, D.W. Griffith, Mae Marsh
Blackboard Jungle, The, Richard Brooks, 1955: Glenn Ford, Sidney Poitier
Blazing Saddles, Mel Brooks, 1974; Mel Brooks
Bonnes Femmes, Les, Claude Chabrol, 1960: Stéphane Audran, Bernadette Lafont
Bonnie and Clyde, Arthur Penn, 1967: Warren Beatty, Faye Dunaway
Boy, The, Oshima Nagisa, 1969: Watanabe Fumio
Bringing Up Baby, Howard Hawks, 1938: Cary Grant, Katharine Hepburn
Brothers and Sisters of the Toda Family, The, Ozu Yasujiro, 1941: Ryu Chishu
Bus Stop, Joshua Logan, 1956: Marilyn Monroe, Don Murray
Casablanca, Michael Curtiz, 1943: Ingrid Bergman, Humphrey Bogart, Sydney Greenstreet, Paul Henreid, Peter Lorre, Claude Rains
Ceremony, The, Oshima Nagisa, 1971
Chagrin et la Pitié, Le, Marcel Ophuls, 1971
Chikamatsu Monogatari, Mizoguchi Kenji, 1954: Hasegawa Kasuo, Kagawa Kyoko

Filmography 205

Chinoise, La, Jean-Luc Godard, 1967: Jean-Pierre Léaud, Anna Wiazomsky
Christopher Strong, Dorothy Arzner, 1933: Katharine Hepburn
Ciel est à Vous, Le, Jean Grémillon, 1943: Charles Vanel
Citizen Kane, Orson Welles, 1941: Joseph Cotten, Agnes Moorehead, Orson Welles
Coming Home, Hal Ashby, 1978: Bruce Dern, Jane Fonda
Coogan's Bluff, Don Siegel, 1968: Clint Eastwood
Corbeau, Le, Henri-Georges Clouzot, 1943: Pierre Fresnay
Cousins, Les, Claude Chabrol, 1958: Claude Brialy
Crime de Monsieur Lange, Le, Jean Renoir, 1936: Jules Berry
Croix de Bois, Les, Raymond Bernard, 1932
Crowd, The, King Vidor, 1928: King Vidor
Dance, Girl, Dance, Dorothy Arzner, 1935: Lucille Ball, Maureen O'Hara
Death by Hanging, Oshima Nagisa, 1968
Deer Hunter, The, Michael Cimino, 1979: Robert de Niro, Meryl Streep
Dentist, The, Leslie Pearce, 1932: W. C. Fields
Deux ou Trois Choses que je sais d'elle, Jean-Luc Godard, 1968: Marina Vlady
Dirty Harry, Don Siegel, 1971: Clint Eastwood
Dodge City, Michael Curtiz, 1939: Errol Flynn
'Dollars' Series, The: *A Fistful of Dollars*, 1964/ *For a Few Dollars More*, 1965/ *The Good, the Bad, and the Ugly*, 1966; dir. Sergio Leone: Clint Eastwood, Gian Maria Volonte
Double Indemnity, Billy Wilder, 1944: Fred MacMurray, Edward G. Robinson, Barbara Stanwyck
East of Eden, Elia Kazan, 1955: James Dean, Burl Ives, Raymond Massey, Jo Van Fleet
Easy Rider, Peter Fonda, 1969: Peter Fonda, Dennis Hopper, Jack Nicholson
El Dorado, Howard Hawks, 1967: Robert Mitchum, John Wayne
Enfants du Paradis, Les, Marcel Carné, 1945: Arletty, Jean-Louis Barrault, Pierre Brasseur, Maria Casarès
Flesh, Paul Morrissey, 1968: Joe Dallesandro
Fond de l'Air est Rouge, Le, Chris Marker, 1978
42nd Street, Lloyd Bacon (dances by Busby Berkeley), 1933: Dick Powell, Ginger Rogers
Français, si vous saviez . . ., André Harris/Alain de Sédouy, 1973
Funny Face, Stanley Donen, 1957: Fred Astaire, Audrey Hepburn
General, The, Buster Keaton, 1926: Buster Keaton
Giant, George Stevens, 1956: James Dean, Rock Hudson, Elizabeth Taylor

Gilda, Charles Vidor, 1946: Rita Hayworth, Glenn Ford
Girl Can't Help It, The, Frank Tashlin, 1956: Tom Ewell, Jayne Mansfield
Godelureaux, Les, Claude Chabrol, 1960: Claude Brialy
Godfather, The (Part 1), Francis Coppola, 1971: Marlon Brando, Diane Keaton, Al Pacino
Gold Diggers of 1933, Mervyn LeRoy (dances by Busby Berkeley), 1933: Joan Blondell, Dick Powell, Ginger Rogers
Gone With the Wind, Victor Fleming (uncredited: George Cukor, Sam Wood), 1939: Clark Gable, Olivia de Havilland, Leslie Howard, Vivien Leigh
Grande Bouffe, La, Marco Ferreri, 1973: Marcello Mastroianni, Philippe Noiret, Michel Piccoli, Ugo Tognazzi
Grande Illusion, La, Jean Renoir, 1937: Marcel Dalio, Pierre Fresnay, Jean Gabin, Erich von Stroheim
Greed, Erich von Stroheim, 1923: Jean Hersholt, Zasu Pitts
Green Berets, The, John Wayne/Ray Kellogg, 1968: Aldo Ray, John Wayne
High Noon, Fred Zinnemann, 1952: Gary Cooper, Grace Kelly
Hiroshima Mon Amour, Alain Resnais, 1959: Emmanuelle Riva
His Girl Friday, Howard Hawks, 1939: Cary Grant, Rosalind Russell
If..., Lindsay Anderson, 1968: Malcolm McDowell
It Happened One Night, Frank Capra, 1934: Claudette Colbert, Clark Gable
I Was a Male War Bride, Howard Hawks, 1949: Cary Grant, Ann Sheridan
Jazz Singer, The, Alan Crosland, 1927: Al Jolson, Myrna Loy
Jour de Fête, Jacques Tati, 1947: Jacques Tati
Jour se Lève, Le, Marcel Carné, 1939: Arletty, Jules Berry, Jean Gabin
Jules et Jim, François Truffaut, 1961: Jeanne Moreau, Henri Serre, Oskar Werner
Kind Hearts and Coronets, Robert Hamer, 1949: Joan Greenwood, Alec Guinness, Dennis Price
King Kong, Ernest Schoedsack/Merian Cooper, 1933: Fay Wray
Lacombe Lucien, Louis Malle, 1974: Pierre Blaise
Lady Vanishes, The, Alfred Hitchcock, 1938: Margaret Lockwood, Basil Radford, Michael Redgrave, Naunton Wayne, Dame May Whitty
Little Big Man, Arthur Penn 1970: Faye Dunaway, Dustin Hoffman
Little Caesar, Mervyn LeRoy, 1930: Edward G. Robinson
Long Goodbye, The, Robert Altman, 1972: Elliott Gould
Lost Weekend, The, Billy Wilder, 1945: Ray Milland, Jane Wyman
Magnificent Seven, The, John Sturges, 1960: Yul Brynner, Steve

Filmography

McQueen, Eli Wallach
Maltese Falcon, The, John Huston, 1941: Mary Astor, Humphrey Bogart, Sydney Greenstreet, Peter Lorre
Manhattan, Woody Allen 1979: Woody Allen, Diane Keaton
Marseillaise, La, Jean Renoir, 1938: Louis Jouvet, Pierre Renoir
Massacre, Alan Crosland, 1935: Richard Barthelmess
Maudite soit la Guerre!, Alfred Machin, 1914
Midnight Cowboy, John Schlesinger, 1969: Dustin Hoffman, Jon Voight
Mildred Pierce, Michael Curtiz, 1945: Ann Blyth, Joan Crawford, Wally Faye, Zachary Scott
Million, Le, René Clair, 1931: Annabella
Mr Deeds Goes to Town, Frank Capra, 1936: Jean Arthur, Gary Cooper
Mr Smith Goes to Washington, Frank Capra, 1939: Jean Arthur, James Stewart
Moi, Pierre Rivière . . ., René Allio, 1975
Monte Carlo, Ernst Lubitsch, 1930: Jeanette MacDonald
Muriel, Alain Resnais, 1963: Jean-Pierre Kerien, Delphine Seyrig
My Ain Folk/My Childhood/My Way Home, Bill Douglas, 1979
Ninotchka, Ernst Lubitsch, 1939: Melvyn Douglas, Greta Garbo
Never Give a Sucker an Even Break, Eddie Cline, 1941: W. C. Fields
Noces Rouges, Les, Claude Chabrol, 1973: Stéphane Audran, Michel Piccoli
North by North-West, Alfred Hitchcock, 1959: Cary Grant, James Mason, Eva-Marie Saint
Numéro Deux, Jean-Luc Godard/Anne-Marie Miéville, 1975
Ohayo/Good Morning, Ozu Yasujiro, 1959: Ryu Chishu
On the Waterfront, Elia Kazan, 1954: Marlon Brando, Eva-Marie Saint, Rod Steiger
Partie de Campagne, Une, Jean Renoir, 1936: Jean Renoir
Passion of Joan of Arc, The, Carl Theodore Dreyer, 1928: Antonin Artaud, Maria Falconetti, Michel Simon
Patrouille de Choc, Claude-Bernard Aubert, 1956
Peeping Tom, Michael Powell, 1959: Carl Boehm
Petit Soldat, Le, Jean-Luc Godard, 1960: Anna Karina
Philadelphia Story, The, George Cukor, 1940: Cary Grant, Katharine Hepburn, James Stewart
Pick-up on South Street, Samuel Fuller, 1953: Thelma Ritter, Richard Widmark
Pierrot le Fou, Jean-Luc Godard 1965: Jean-Paul Belmondo, Anna Karina
Play It Again, Sam, Herbert Ross, 1972: Woody Allen
Portrait of Madame Yuki, Mizoguchi Kenji, 1950

Psycho, Alfred Hitchcock, 1960: Janet Leigh, Anthony Perkins
Public Enemy, William Wellman, 1931: James Cagney, Jean Harlow
Quai des Brumes, Marcel Carné, 1938: Jean Gabin, Michèle Morgan, Michel Simon
Queen Kelly, Erich von Stroheim, 1928: Gloria Swanson, Erich von Stroheim
Que la Bête Meure!, Claude Chabrol, 1969: Jean Yanne
Rashomon, Kurosawa Akira, 1950: Kyo Muchiko, Mifune Toshiro
Rebel Without a Cause, Elia Kazan, 1955: James Dean, Sal Mineo, Natalie Wood
Red Dust, Victor Fleming, 1932: Mary Astor, Clark Gable, Jean Harlow
Red River, Howard Hawks, 1948: Walter Brennan, Montgomery Clift, Joanne Dru, John Wayne
Règle du Jeu, La, Jean Renoir, 1939: Marcel Dalio, Nora Gregor, Gaston Modot, Jean Renoir
Religieuse, La, Jacques Rivette, 1965: Anna Karina
Rio Bravo, Howard Hawks, 1959: Walter Brennan, Angie Dickinson, Dean Martin, Rick Nelson, John Wayne
Rio Grande, John Ford, 1950: Maureen O'Hara, John Wayne
Ruggles of Red Gap, Leo McCarey, 1938: Charles Laughton, Zasu Pitts
Sanjuro, Kurosawa Akira, 1962: Mifune Toshiro
Sansho Dayu, Mizoguchi Kenji, 1954: Shindo Fitaro, Tanaka Kinuyo
Scarface, Howard Hawks, 1932: Paul Muni, George Raft
Scarlet Empress, The, Josef von Sternberg, 1934: Marlene Dietrich
Searchers, The, John Ford, 1956: Jeffrey Hunter, John Wayne, Natalie Wood
Semaine de Vacances, Une, Bertrand Tavernier, 1980: Nathalie Baye, Philippe Noiret
Seven Samurai, The, Kurosawa Akira, 1954: Mifune Toshiro
Shanghai Express, Josef von Sternberg, 1932: Clive Brook, Marlene Dietrich
Shock Corridor, Samuel Fuller, 1963: Peter Breck
Singin' in the Rain, Stanley Donen/Gene Kelly, 1952. Cyd Charisse, Jean Hagen, Gene Kelly, Donald O'Connor, Debbie Reynolds
Sous les Toits de Paris, René Clair, 1930: Paula Illery, Albert Préjean
Stagecoach, John Ford, 1939: Claire Trevor, John Wayne
Story of the Last Chrysanthemums, The, Mizoguchi Kenji, 1939
Sunset Boulevard, Billy Wilder, 1950: William Holden, Gloria Swanson, Erich von Stroheim
Swing Time, George Stevens, 1936: Fred Astaire, Ginger Rogers
Tarnished Angels, The, Douglas Sirk, 1952: Rock Hudson, Dorothy Malone, Robert Stack

Filmography

Taxi Driver, Martin Scorsese, 1976: Robert de Niro, Jodie Foster
There was a Father, Ozu Yasujiro, 1942
They Live by Night, Nicholas Ray, 1948: Farley Granger, Cathy O'Donnell
Three Little Pigs, The, Walt Disney, 1933
Tirez sur le Pianiste, François Truffaut, 1960: Charles Aznavour
To Have and Have Not, Howard Hawks, 1945: Lauren Bacall, Humphrey Bogart, Walter Brennan
Tokyo Story, Ozu Yasujiro, 1953: Ryu Chishu
Toni, Jean Renoir, 1935: Claude Renoir
Trash, Paul Morrissey, 1970: Joe Dallesandro
Twentieth Century, Howard Hawks, 1934: John Barrymore, Carole Lombard
Ugetsu Monogatari, Mizoguchi Kenji, 1953: Kyo Machiko
Underworld USA, Samuel Fuller, 1961: Cliff Robertson
Vie est à Nous, La, Jean Renoir, 1936: Claude Renoir
Visiteurs du Soir, Les, Marcel Carné, 1942: Arletty, Jules Berry
Vivre sa Vie, Jean-Luc Godard, 1962: Anna Karina
Voyage dans la Lune, Georges Méliès, 1902
Weekend, Jean-Luc Godard, 1967: Mireille Darc, Jean Yanne
Whisky Galore, Alexander Mackendrick, 1949: Joan Greenwood
White Heat, Raoul Walsh, 1949: James Cagney, Virginia Mayo
Wild One, The, Laszlo Benedek, 1953: Marlon Brando, Lee Marvin
Written on the Wind, Douglas Sirk, 1956: Lauren Bacall, Rock Hudson, Dorothy Malone, Robert Stack
Yojimbo, Kurosawa Akira, 1961: Mifune Toshiro
Young Mr Lincoln, John Ford, 1939: Henry Fonda
You Only Live Once, Fritz Lang, 1937: Henry Fonda, Sylvia Sidney
Zéro de Conduite, Jean Vigo, 1933: Henri Storck

Index

A Bout de Souffle 42, 140, 141
Académie Française 122
Across the Pacific 55
Adam's Rib 33
Afrique 50 138
Age d'Or, L' 119,122
Ai No Corrida (Empire of the Senses) 177, 193, 194
Air Force 55
Alamo, The 3
Algeria 111, 139, 143-4, 151
All About Eve 77-8, 80, 175
Allen, Woody 98
Allio, René 117
All That Heaven Allows 86
Almereyda, Miguel 122
Alphaville 144, 145
Altman, Robert 18, 63
Amants, Les 142
American Graffiti 4, 85, 98, 113
Anderson, Joseph L. 176, 178, 179
Anderson, Lindsay 122, 169
Angels With Dirty Faces 48-50 *passim*
Anger, Kenneth 61
Annie Hall 98
A Nous la Liberté 120-21, 124
Apocalypse Now 11, 101, 104-7 *passim*
Aragon, Louis 122
Arbuckle, Fatty 63
Army Film Services 116
Arroseur Arrosé, L' 115
Arthur, Jean 31, 38
Arzner, Dorothy 38-9
Ashby, Hal 102
Astaire, Fred 5, 81-2
Astor, Mary 56-8 *passim*
Atalante, L' 123
Auden, W. H. 179
Autumn Afternoon, An 180-81

Bacall, Lauren 62
Backus, Jim 82
Balazs, Bela 22
Ball, Lucille 39
Balzac, Honoré de 19, 119
Bara, Theda 25
Barbour, Alan G. 85
Barr, Charles 163, 165, 168-9
Barrès, Maurice 134
Barrow, Clyde 92
Barrymore, John 63
Bataille, Georges 156, 167
Baudrillard, Jean 173
Baxter, Anne 77, 176
Bazin, André 68-71 *passim*, 119, 189n.
Beatty, Warren 92
Beaumarchais, Pierre-Auguste 137
Belmondo, Jean-Paul 140, 145
Bergman, Andrew 30-31
Bergman, Ingrid 62
Berkeley, Busby 29, 30, 36
Bernard, Raymond 117
Best Years of Our Lives, The 68-70
Beyond the Blue Horizon 27
Biches, Les 141
Big Bad Wolves 69-70
Big Parade, The 25
Big Sleep, The 56
Bill of Rights 26
Birth of a Nation, The 21, 41
Blackboard Jungle, The 85
Blazing Saddles 183
Blyth, Ann 59
Boehm, Carl 157
Boetticher, Bud 16
Bogart, Humphrey 14, 17, 56-7, 62-5, 85, 124
Bogle, Donald 40
Bohème, La 119
Bonnes Femmes, Les 141
Bonnie and Clyde 92-4 *passim*
Borde, Raymond 55
Borges, Jorge Luis 178n.
Boy, The 192
Brahms, Johannes 141

Index

Brando, Marlon 87, 92, 99, 104
Braudy, Leo 118, 129
Brewer, Roy 73
Bringing Up Baby 33-36 *passim*, 73
British Broadcasting Corporation 159-60
British Film Institute 5n., 15, 160-61
Brook, Clive 36
Brooks, Mel 183
Brothers and Sisters of the Toda Family, The 179
'Bullet in the Head : Vietnam Remembered, A' 103
Buñuel, Luis 155
Burch, Noël 171, 182, 189, 191
Bus Stop 95

Cagney, James 14, 48, 87
Cahiers du Cinéma 7, 148, 161
Camus, Albert 126
Cannes Film Festival 148
Capitalism and Freedom 13
Capone, Al 47
Capra, Frank 31, 32
Carette, Julien 124, 135
Carné, Marcel 119, 123-6 *passim*, 128
Carson, Jack 59
Casablanca 4, 17, 60-61, 65, 98, 134
censorship 4, 14, 19, 55, 116, 121, 124-5, 128, 131-2, 138-9, 155
Ceremony, The 192
Chabrol, Claude 128, 141, 142
Chagrin et la Pitié, Le 149, 150
Chaplin, Charlie 19, 24, 123
Chapman, Graham 157n.
Chaumeton, Étienne 55
Chéri 143
Chikamatsu Monogatari 9
Chinoise, La 146-8
Christie, Ian 158
Christopher Strong 38-9
CIA 89
Ciel est à Vous, Le 127-9
Cinéma et Idéologie 114
Cinéthique 148
Citizen Kane 3, 60, 65-8, 69, 105-6, 137
Clair, René 119-23 *passim*, 127, 129, 140
Clarke, Mae 49
Clift, Montgomery 71-2, 82
Clouzot, Henri-Georges 129-30
Colbert, Claudette 29
Coleman, John 160
Colette 143
Combs, Richard 102
Coming Home 101-2, 106
Communism 13, 20, 26, 67, 73, 74-6, 87, 88-9, 102-4, 121-2, 123, 127, 130-31, 146-7
Conference of Studio Unions 73
Confidential 61
Conrad, Joseph 19n., 104, 143
Coogan's Bluff 91, 94-7 *passim*
Cook, Jim 63n.
Cooper, Gary 20, 31, 38, 75-6
Coppola, Francis Ford 11, 99, 105
Corbeau, Le 128
Cotteret, Jean-Marie 110

Courtade, Francis 125, 127, 131
Cousins, Les 142
Coward, Ros 160n.
Crawford, Joan 58, 61
Crime de Monsieur Lange, Le 129-30
Croix de Bois, Les 117
Crosland, Alan 46
Crowd, The 25
Cukor, George 32, 51
Curtiz, Michael 48, 60-61, 134

Daily Telegraph 158
Dance, Girl, Dance 39
Darc, Mireille 147
Daughters of the American Revolution 92
Davies, Marian 65
Davis, Bette 61, 77-9, 176
Dead End Kids 50
Dean, James 79, 82, 85
Death by Hanging 193
Debray, Régis 109-10, 122, 128, 151
Deer Hunter, The 101-4 *passim*, 106
Demaria, M. 116
Dentist, The 22
Depression 20-21, 28-34 *passim*, 37, 41-2, 55, 60, 77, 84, 92-3, 141
Deux ou Trois Choses que je sais d'elle 145, 147, 149
Dewey, John 73
Didion, Joan 98
Dietrich, Marlene 29, 33, 36-7
Dirty Harry 94-5
Disney, Walt 27, 30
Dixon, Campbell 158
Dodge City 43, 91
Dollars series 91
Dorgelès, Roland 117
Dostoevsky, Fyodor 19n.
Double Indemnity 56-9
Douglas, Bill 154
Douglas, Lord Alfred 157
Dreyer, Carl 22-3
Dreyfus, Alfred 133
Dru, Joanne 72
Durgnat, Raymond 83, 90, 155
Dyer, Peter John 73
Dyer, Richard 5n.

Ealing Studios 169
Earley, Steven C. 41
Easthope, Antony 15
East of Eden 82
Eastwood, Clint 94-5
Easy Rider 4, 93-4
Éclair film company 116
Eclipse film company 116
Edinburgh Festival 5
Eighteenth Brumaire of Louis Bonaparte, The 110
Eisenstein, Sergei 68-9, 174n.
El Dorado 76
Electric Cinema Club 5
Eliot, T. S. 104
Ellis, John 160n.
Empire Marketing Board 153
Enfants du Paradis, Les 126

Evans, Robert 99
Évian agreement 111

Fairbanks, Douglas 25
Fascism 56, 112, 117, 121-31 *passim*, 143, 144
Fassbinder, Rainer Werner 86
FBI 48
Ferreri, Marco 183
Ferro, Marc 115, 150-51
Fetchit, Stepin 40
Fields, W. C. 22, 47, 64
Fifth Amendment 75
Film, Le 116
film noir 11, 16-17, 18, 41, 53-68, 87
First World War 25, 28, 42, 112, 113, 116, 117, 132, 133, 134, 140-41, 149
Fistful of Dollars, A 172
Fitzgerald, F. Scott 47
Flaubert, Gustave 143, 159
Fleming, Victor 51
Flesh 94
Flynn, Errol 43, 61, 63-4, 91
Fonda, Jane 101
Fonda, Peter 93
Fond de l'Air est Rouge, Le 113
Fontaine, Joan 175
Ford, Glenn 57
Ford, John 5, 7, 42-3, 95, 102, 106, 161
42nd Street 28-9
Foucault, Michel 177
Fountainhead, The 12n.
Four Serious Songs 141
Français, si vous saviez . . . 111, 150
Frazer, Sir James 104
Freed, Arthur 41, 80
Fresnay, Pierre 132
Freud, Sigmund 33, 49, 66, 67, 86, 121, 148, 160, 161
Friedman, Milton 13-14, 16, 19, 74
Fuller, Samuel 4, 47, 87-8, 99
Funny Face 81

Gabin, Jean 3, 124-5, 128, 132-3
Gable, Clark 29, 36, 51-3
Gang, Martin 75
gangster movies 2, 18, 19-20, 41-50, 55, 70, 86-7
Garbo, Greta 65, 85
Garland, Judy 61
de Gaulle, General 110-12, 131, 145, 149-52 *passim*
Gaumont Studios 116
General, The 23
Genre 15
Giant 82
Gilbert, John 63
Gilda 56-7
Giraudoux, Jean 132
Girl Can't Help It, The 83, 85, 156
Giscard d'Estaing, Valery 149
Godard : Images, Sounds, Politics 149
Godard, Jean-Luc 42n., 114, 140, 143-5, 148-9, 151, 161, 177
Godelureaux, Les 142
Godfather, The 18, 20, 98-100

Gold Diggers of 1933 20, 28, 92
Gold Diggers series 18, 28-9, 39
Goldmann, Annie 119
Gone with the Wind 50-53
'Good Morning' (song) 80
Goon Shows 170
Gould, Elliot 63
Grande Bouffe, La 183
Grande Illusion, La 34, 112-13, 131, 134-5, 149
Grant, Cary 32-6, 38, 72-3, 89
Greed 22-3, 25
Green Berets, The 3, 7, 101, 105
Greenberg, Joel 57
Greenstreet, Sydney 56
Greenwood, Joan 163, 166
Gregor, Nora 135
Griffith, D. W. 21, 23, 114
Gruber, Frank 42
Guinness, Sir Alec 162

Hagen, Jean 80-81
Hamer, Robert 168, 170
Hammer Films 12
Hancock, Tony 155n.
Harlow, Jean 36-7
Harris, André 111, 151, 152
Harvey, Sylvia 146
Haskell, Molly 58
Hawks, Howard 7, 32, 38, 55, 70, 76
Hays Production Code 14, 29, 31, 48, 55, 63, 73, 85
Hayworth, Rita 56-8 *passim*, 98
Hearst, William Randolph 65
Heath, Stephen 193
Hemingway, Ernest 102
Hemingway, Mariel 98
Henreid, Paul 62, 134
Hepburn, Audrey 81-2
Hepburn, Katharine 32-6 *passim*, 38
Higham, Charles 57
High Noon 28, 75-7 *passim*, 87, 91
Hill, Derek 158, 161
Hiroshima Mon Amour 142
His Girl Friday 38
Histoire de l'Oeil 156
Historical Novel, The 19n.
Hitchcock, Sir Alfred 15, 83, 89-90, 155, 161, 175
Hobson, Valerie 166
Hoffman, Dustin 94
Holden, William 78-9
Holly, Buddy 114
Hollywood Babylon 61
Hollywood Communist Party 73
Homer 5
Hopper, Dennis 93
Hopper, Hedda 61
House Committee on Un-American Activities (HUAC) 73, 75, 87, 138
Houston, Penelope 177, 178
Howard, Leslie 52
How to Read Donald Duck 30n.
Hudson, Rock 86
Hughes, Howard 66
Humanité, L' 131

Index

Hunter, Paul 53
Huston, John 55

IDHEC film-school 139
If... 122
Images of Alcoholism 63n.
International Alliance of Theatrical Staff Employees (IA) 73
Introduction to American Movies, An 41
It Happened One Night 29, 90
I Was a Male War Bride 33, 70, 72
I Was Born, But... 182

Jacobs, Lewis 20, 24
James, Sid 155
Japanese Film, The 178
Jazz Singer, The 26
Jeanson, Henri 146
Jeanson, Juliette 145
Jour de Fête 142
Jour se Lève, Le 3, 123-4, 125
Jules et Jim 140, 141

Kabuki theatre 191
Karina, Anna 143, 145
Kazan, Elia 47, 87
Keaton, Buster 23-4, 63, 78
Keaton, Diane 98-9
Kelly, Gene 79-80
Kelly, Grace 76-7
Kennedy, President John 91
Kerouac, Jack 92
Keystone Cops 23
Kind Hearts and Coronets 2, 157, 162-3, 165, 167-70 *passim*
King Kong 30-31, 36, 105, 116, 123
Kinnear, Roy 155
Kinsey Report 85
Kurosawa, Akira 9, 172-3, 176-7, 183, 185-6

Lacan, Jacques 161
de Laclos, Choderlos 168
Lacombe Lucien 150
Lady Chatterley's Lover 154-5, 160
Lady Vanishes, The 154-5
Lang, Fritz 55
Langlois, Henri 148
Language and Materialism 160n.
LaRue, Danny 167
Laurent, Jacqueline 124
Lawrence, D. H. 159
Leavis, F. R. 159
Lebel, Jean-Patrick 114
Legion of Decency 29, 121, 155
Leigh, Vivien 51
Lejeune, C. A. 158
Le Roy, Baby 29
Leutrat, Jean-Louis 15
Lewington, Mike 63n.
Lewton, Val 41
Liaisons Dangereuses, Les 168
Liberty 53
Lindbergh, Charles 47
Little Big Man 46, 91
Little Caesar 47, 49

Lloyd, Harold 24
Long Goodbye, The 63
Lorre, Peter 40, 56, 79
Lost Weekend, The 63
Lovell, Alan 95
Lubitsch, Ernst 27
Lucas, George 113
Lukacs, Georg 19n.
Lumière brothers 114-15

MacDonald, Dwight 67
MacDonald, Jeanette 27
Machin, Alfred 116
MacMurray, Fred 56-7
Madame Bovary 143
Mafia 18, 99
Magnificent Seven, The 43, 172
Malle, Louis 142
Malone, Dorothy 86
Malraux, André 102
Maltese Falcon, The 40, 55-7, 62
Manhattan 98
Mankiewicz, Herman 66
Mansfield, Jayne 83, 97
Mariage de Figaro, Le 137
Marker, Chris 113
Marseillaise, La 3, 112-13, 133, 149
Martin, Dean 76
Martin, Millicent 155
Marx, Karl 17, 19, 110, 148, 160
Massacre 46
Mattelart, Armand 30
Maudite soit la Guerre! 116
du Maurier, Daphne 175
May 68 and Film Culture 148n.
McArthur, Colin 18, 42, 48
McCarthy, Senator Joseph 2, 13-14, 26, 75-6, 85, 89, 91, 121
McDaniel, Hattie 40
McGill, Donald 156, 162, 169
Méliès, Georges 115
Mellen, Joan, 70, 185, 187
Methot, Mayo 62
Metro-Goldwyn-Mayer 14, 28, 41-2, 51, 55, 79-80
Metz, Christian 30, 161, 189
Midnight Cowboy 94
Miéville, Anne-Marie 149
Mifune, Toshiro 9, 184, 187
Mildred Pierce 3, 58-60
Million, Le 119-20
Mineo, Sal 82
Mishima, Yukio 175
Mitchell, Margaret 51
Mitchum, Robert 76
Mizoguchi, Kenji 3, 175-7 *passim*, 183, 185, 188-92
Moi, Pierre Rivière 117
'Momism' 20, 83-4
Monogram Pictures 16, 42
Monroe, Marilyn 83, 96, 97
Monte Carlo 27
Monthly Film Bulletin 158
Monty Python 157
Morecambe, Eric 155n.
Moreau, Jeanne 141-2

Moreau, René 110
Morgan, Michèle 124
Morrissey, Paul 94
Motion Picture Alliance for the Preservation of American Ideals 85
Motion Picture Producers and Distributors of America Inc. 29
Movie-Made America 33
Mr Deeds Goes to Town 31-2, 38, 96
Mr Smith Goes to Washington 31
Munich Agreement 123
Muriel 142, 144
Murray, Don 95
My Ain Folk 154
My Childhood 154
My Way Home 154

National Association for the Advancement of Coloured People 21
National Film Theatre 159
Nazism 126, 128
Neale, Stephen 15
Never Give a Sucker an Even Break 64
'New Deal' 30, 33, 42, 91, 96
New Statesman 160
Nicholson, Jack 93
Ninotchka 85
de Niro, Robert 96
Noces Rouges, Les 141-2
Noh theatre 174, 191
North by North-West 89-90
Numéro Deux 149

O'Brien, Pat 49-50
Observer 158
Occupation (France) 126-8, 134, 142, 149-50
O'Connor, Donald 80
Odyssey 5
O'Hara, Maureen 39
Ohayo$good Morning 179, 182
On the Waterfront 47, 87
ORTF 114, 138
Oshima, Nagisa 161, 176-7, 188, 192-3
Ozu, Yasujiro 3, 175-85 *passim*, 192

Pacino, Al 99
Pagnol, Marcel 118, 124
Pakula, Alan 18
Panorama du Film Noir Américain 55
Paramount Pictures 41
Paris Cinémathèque 148
Parsons, Louella 61, 78
Partie de Campagne, Une 112
Passion of Joan of Arc, The 22-3
Pathé 116
Patrouille de Choc 139
Pearl Harbor 53, 60
Peeping Tom 157-62 *passim*, 170
Perkins, Anthony 83
Peste, La 126
Pétain, Marshal 116, 127-8
Petit Soldat, Le 139, 143-4
Philadelphia Story, The 33, 34-6

Pickford, Mary 25
Pick-Up on South Street 20
Pierrot le Fou 145
Play It Again, Sam 98
Popcorn Venus 37
Portrait of Madame Yuki 188, 191
Powell, Michael 157-60 *passim*, 170
Prédal, René 113, 117, 126
Presley, Elvis 84
Pressburger, Emeric 159-60
Prévert, Jacques 123-4, 126
Price, Dennis 162, 165
Prohibition 18, 47, 86-7
Psycho 20, 83-5, 88
Public Enemy, The 48-9, 83, 87
Pulitzer Prize 88
Pym, John 103

Quai des Brumes 123-5 *passim*
Que la Bête Meure! 141
Queen Kelly 78
Question Oshima, The 193

Radford, Basil 154
Rains, Claude 62
Rand, Ayn 12, 19, 74
Rappe, Virginia 63
Rashomon 183-5 *passim*
Rebecca (book and film) 175
Rebel Without a Cause 82, 85
Recherches sur le Vocabulaire du Général de Gaulle 110
Red Dust 36
Red River 7, 20, 70-72, 82, 90, 91, 113, 129
Règle du Jeu, La 3, 112, 114, 134-5, 137, 142, 167, 190
Religieuse, La 138, 155
Rendez-Vous des Quais, Le 138
Renoir, Jean 3, 34, 69, 112, 118-19, 125-31 *passim*, 133-4, 137, 151, 189n.
Resistance movement 126-7, 134, 149-51
Resnais, Alain 142, 144, 151
Reynolds, Debbie 80
Rhode, Eric 189, 191
Richie, Donald 176, 178, 179
Rio Bravo 76, 91
Rio Grande 7
Rise of the American Film, The 20
RKO Radio Pictures 41-2
Robinson, David 158
Robinson, Edward G. 48, 57
Roosevelt, President F. D. 30, 41
Rosen, Marjorie 37, 77
Ruggles of Red Gap 26
Russell, Ken 96n.
Russell, Rosalind 38

de Sade, Marquis 158, 167
Saint, Eva-Marie 89
Sanders, George 77
Sanjuro 172, 185
Sansho Dayu 188, 192
Scarface 47
Scarlet Empress, The 29
Schrader, Paul 178, 190
Scott, Zachary 59

Index

Screen 160-62 *passim*
Screen Education 15
Screen Guide for Americans 14
Scrutiny 159
Searchers, The 5-9, 90
Seberg, Jean 140
Second Coming, The 70
Second World War 3, 14, 17, 41, 54, 70, 113-114, 119, 123-5, 132, 134, 137, 172-3, 175, 188
de Sédouy, Alain 111, 151, 152
Selznick, David 51
Semaine de Vacances, Une 148
Sennett, Mack 23
Seven Samurai, The 43n., 172, 185, 187
Seven Year Itch, The 96n.
Shakespeare, William 190
Shanghai Express 29, 36
Sheen, Martin 104
Shepherd, Cybill 97
Sheridan, Ann 32, 72
Shock Corridor 88-9, 97
Siegel, Donald 94-5
Siegfried et le Limousin 132
Signs and Meaning in the Cinema 69n.
Singin' in the Rain 18, 79-81
Sir Henry at Rawlinson End 170n.
Sirk, Douglas 4-5, 16, 17, 85-6, 102
Six-Guns and Society 43, 46
Sklar, Robert 33
Smith, Adam 19
Society for Education in Film and Television 160, 161
Sorlin, Pierre 5
Sous les Toits de Paris 120
Soustelle, Jacques 151
Stack, Robert 86
Stagecoach 44-6, 47, 87, 90
Stanshall, Vivien 170n.
Stanwyck, Barbara 56-8
Stars 5n.
Stavisky affair 122
von Sternberg, Josef 37
Stevens, George 11
Stewart, James 31-2, 34
Story of the Last Chrysanthemums, The 9, 188-9
von Stroheim, Erich 22, 78-9, 132
Studies in European Realism 19n.
Sturges, Preston 32
Sunset Boulevard 77, 78-9
Swanson, Gloria 78-9
Swing Time 11
Système des Objets, Le 173

Tallents, Sir Stephen 153
'Tara's Theme' 53
Tarnished Angels, The 86
Tati, Jacques 142
Taxi Driver 96-7
Taylor, Elizabeth 79
Temple, Shirley 29
Tennessee Valley Authority 42
There was a Father 178
They Live by Night 92
Thomas, Parnell 75

Thomson, David 50, 95
Three Little Pigs (characters) 30
Till Death Us Do Part 186
Tirez sur le Pianiste 140
To Have and Have Not 62
Tokyo Story 177-82 *passim*
Toland, Gregg 68-9
Tolstoy, Leo 78, 192
Tommy 96n.
Toms, Coons, Mulattoes, Mammies, and Bucks 40n.
Toni 118
Tracy, Spencer 33, 36
Trash 94
Travolta, John 79
Tribune 158
Truffaut, François 140
Twentieth Century 63
Twentieth-Century Fox Film Corporation 55

Ugetsu Monogatari 177, 189-92
Under Western Eyes 143
Underworld USA 47, 87, 88-9
United Artists Corporation 42
Universal Pictures 16, 41-2, 55, 85-6

Vacances de M. Hulot, Les 142
Valentino, Rudolph 26
Vautier, René 138
Venice Film Festival (1951) 183 (1959) 134
Verleugnung 161
Vickers, Martha 56
Vidor, King 25
Vie est à Nous, La 112, 130-131
Vigo, Jean 119, 121-2
Visiteurs du Soir, Les 126
Vivre sa Vie 145
Voight, Jon 94
Voyage dans la Lune 115

Wage Earners Committee 74
Wall Street 30
Wanger, Walter 53
War and Peace 78
Warhol, Andy 94
Warner Brothers 28, 36, 41-2, 50, 55
Warner, Jack 28, 41
Warshow, Robert 70-71
Watergate 44n., 91, 95, 101
Wayne, John 3, 5-9, 16, 20, 43, 44n. 70-72, 75-7, 90-91, 100, 124
Wayne, Naunton 154
Weber, Eugen 133
Weekend 147
Welles, Orson 41, 64, 65-6, 75, 137, 189n.
Wellman, William 48
'We're in the Money' (song) 20, 28
We're in the Money (book) 30
West, Mae, 65, 155
'Westerner, The' 70
Westerns 1, 3, 6-9 *passim*, 11-12, 14, 16, 18-20, 22, 70-77, 86, 87, 90-92, 95, 100, 102, 113, 172-3
Whisky Galore 154
Whistler, James 152, 157

White Heat 20
Whitehouse, Mrs Mary 159
Whitty, Dame May 154, 166
Wilde, Oscar 153, 157, 169
Wilder, Billy 55
Wild One, The 92
Williams Committee on Obscenity and Film Censorship 54
Wise, Ernie 155n.
Wollen, Peter 69n.
Wood, Michael 80
Wood, Natalie 5
Wood, Sam 5
Work of Dorothy Arzner – Towards a Feminist Cinema, The 39

Wray, Fay 30, 36
Wright, Will 43-4, 46, 76, 185
Written on the Wind 86
Wyler, William 68
Wyman, Jane 86

Yanne, Jean 147
Yeats, W. B. 70
Yojimbo 172, 185, 187
Young Mr Lincoln 161
'Young, Willing, and Healthy' 29
You Only Live Once 92, 124

Zéro de Conduite 4, 119-23 *passim*